THE RED KITE

CALIOLOGISTS' SERIES No.1

Second Edition

M.J. DAWSON

ORIEL STRINGER

THE CALIOLOGISTS' SERIES

No 1:	THE RED KITE 2nd Edition	0 948122 06 4
No 2:	THE BLACK-THROATED DIVER	0 948122 01 3
No 3:	CRYPTANALYSIS OF ORNITHOLOGICAL LITERATURE	0 948122 04 8
No 4:	THE PEREGRINES of the LAKE DISTRICT	0 948122 02 1
No 5:	H. J. K. BURNE OOLOGIST	0 948122 05 6
No 6:	THE GOLDEN EAGLE	0 948122 03 X
No 7:	THE OSPREY	0 948122 07 2

ORIEL STRINGER

BOOKS FOR THE FIELD ORNITHOLOGIST

BRIGHTON

BN1 5ND

RED KITE
by Bryan Hanlon
ORIGINAL: WATERCOLOUR 12" x 9"

THE RED KITE
(CALIOLOGISTS' SERIES No. 1)

Second Edition

To
Jane Beaty

THE CALIOLOGISTS' SERIES No: 1

THE RED KITE: *Milvus milvus*

Nesting Sites, Old and New, in Great Britain

M.J. DAWSON

THE RED KITE

Typeset in Baskerville
by Blackmore Typesetting Services · Brighton

SECOND EDITION
ISBN 0 948122 06 4

Drawings by: Tessa Dowland

Published by ORIEL STRINGER
BRIGHTON
ENGLAND

CONTENTS

ACKNOWLEDGMENTS

The author would like to acknowledge the help and advice, particularly in the field, of some members of the Jourdain Society. The Jourdain Society is the only body of ornithologists specialising in the study of the breeding biology of birds of the western palaearctic region and of birds in the British Isles in particular. Many members specialise in one particular species or group of species and their knowledge in this field is probably greater than all other ornithologists in Europe, or in the world.

Particular thanks are extended to Graham Bowes, G.W.R.D., Darren Hughes and T.D., all of whom helped tremendously in the field with their knowledge and untiring energy.

PREFACE

Repeated demands for the Red Kite have necessitated this second edition. Most of the contents are identical but the nesting sites have been brought up-to-date and a short description is given on each.

Since the First Edition was published, the number of Kites has increased enormously. Each year produces better results. This is in spite of more and more molestation by self-appointed 'protectionists'. Nest are repeatedly climbed to, young birds are having 'tags' affixed to their wings and we now read that the eggs of Kites are to be removed from the nests and put into incubators. If the birds were left alone, they would increase much more quickly.

These 'protectionists' intimated that the publication of the Red Kite would be harmful to these birds. However, the opposite is the result, the birds are more numerous than in the last one hundred years.

Three books have so far been published on the finding of birds' nests. The first was Gosnell's The Science of Birdnesting in 1947. The second book was Finding Nests by Campbell, 1953. This, he writes, is entirely ethical, as he is one of the bird-protectors. The third volume is A Field Guide to Birds' Nests by the above Campbell and James Ferguson-Lees. This gives full details on all the nesting birds of Great Britain, where to look and how to seek them out. Again, this is quite legitimate, as the authors are protectionists. Now we have the books in the Caliologists' Series, giving exactly the same information but each book on one species and giving more information. It is no wonder that these books have been described as the most useful bird books in the world to bird-watchers. They are intended to help all those field men who study birds, particularly in a scientific manner. Certes, it is nonsense to imply that these books will encourage persons to take eggs.

All egg collectors already know the information in these books and it is mainly through them that the information is available.

The latest Wildlife & Countryside Act 1981, which came into force on 20th September 1982, states that a specially protected bird must not be disturbed intentionally at or near its nest while the latter is being built or contains eggs or unflown young. Nor must the young be disturbed while they are dependant on their parents. Watching nests from a distance will not, in any way, disturb the birds. All ornithologists should read the Act and become acquainted with the contents.

Wild birds do not belong to the person over whose land they fly or nest on. Game birds have their own Acts to protect them.

The finding of nests is now a pursuit carried out by many persons, not only by professional ornithologists but by people who are interested in seeing the birds in their habitat.

CHAPTER 1

DISTRIBUTION of the Red Kite in WALES up to Circa 1820

In order to ascertain the numbers and distribution of any particular species of bird say, over two hundred years ago, is not an easy task for the ornithologist. So little was written on the subject in an accurate and ornithological sense that any conclusions reached must be mainly conjecture. Examining all the available early ornithological books, really very little has been written upon the Kite. Nineteenth century books, particularly of the latter five decades, mention that the species was once quite a common bird but had since declined in numbers to a very small remnant. E.T. Booth obtained a pair of adult Kites in Perthshire in 1877 and they were, no doubt, one of the few pairs to survive in North Britain. In his Rough Notes (a description which belittles these three mighty volumes) Booth mentions that there were three or four pairs of Kites within a distance of twenty to twenty-five miles. Another pair of which he knew was twenty-five miles up the glen. Booth states that forty years before (i.e. about 1820) Kites were common in the Highlands.

Comparing late authors leads one to state that few of these ornithologists knew of the kite personally and that the meagre treatment of the bird has been copied down through the years nearly word for word. A few odd lines do appear at intervals and the now well-known quotation from Shakespeare on the Kite and 'lesser linen' does not go into detail. Reference to Kites in London in conjunction with present day knowledge of the status and habits of European birds leads us to the conclusion that these were Black Kites. The Red Kite was never very fond of decidedly urban areas, whereas the other species is quite common in such places, in particular in parts of North Africa. Nevertheless, Red Kites could have been well-distributed over the greater part of the British Isles. The only reliable source of bird species around the time of the first printed book is contained in works on field sports. Accounts of Kite-hawking suggest that these birds must have been reasonably common, at least in certain areas, otherwise falcons would not have been trained at that quarry. It was the most superb and most difficult of all forms of falconry and a Peregrine entered at Kite would not have been flown at any other quarry, lest it lessen its tremendous powers of flight, or, at least, refuse the more difficult species. This means that falconers could expect to find a Kite whenever they wished to indulge in this particular branch of the sport. In connection with this, such Kite-hawking could only be carried out in vast open and usually very flat countryside. Again, this leads us to suppose that if the Kite were fairly common in open country, then they were even more common in the wooded hillsides which would, even then, be their main breeding areas.

Hewitson, writing in 1846, remarks that the Kite was rather local. It was confined, in England, to some of the large woods of the Midland and Southern counties and the wilder districts of the North. It was not uncommon in some parts of Wales and Scotland; it was plentiful in Aberdeenshire. It occurred in the vicinity of Loch Katrine and had been seen at Loch Awe. He saw two or three together at the large woods in the neighbourhood of Alconbury Hill, where it always bred.

Wolley, the expert ornithologist, had eggs in his collection from the 1840's and 1850's from the following localities:

A clutch of two eggs taken in 1843 in Monks Wood, and he states that observers could remember that kites were once plentiful in the neighbourhood.

1 egg taken in the New Forest in 1850.

1 egg taken near Alconbury Hill, Huntingdon and he says the birds were becoming rare then, in 1844.

2 eggs taken by Lewis Dunbar in Scotland from the area between Loch Garten and the River Nethy in 1850.

A c/3 from Lincolnshire taken in 1853.
1 egg from North Devon taken in 1855.
On the 14th May 1856 a c/3 from Ballington Wood, Lincolnshire, and, in the same year, on 17th May 1856 an egg from a nest in the Grey Clift Wood, Lincolnshire.
From Branston, Lincolnshire, a c/1 taken on the 16th May 1857.

This shows that the Kite was quite widespread in those years, even though not common. The references to Lincolnshire are interesting, in that this was one of the main Kite-hawking areas.

After many years of studying birds in general, one gathers a particular interest in certain species or in birds of a particular habitat. Having visited various places in Great Britain, the author evolved a special liking for Wales, particularly Radnorshire. Ravens and Buzzards were the most interesting, nesting as they do, in the wilder parts of the terrain. One day, when in the area of Rhandirmwyn, the author met a most congenial American and his charming wife. They had been waiting on the Pen y Darren hill for over an hour, hoping to see a Red Kite in the air but had no sight of any such birds. The author took these two ornithologists to a number of Kite sites and, over several days, these two visitors were able to watch eight pairs of Kites at close quarters. To show their appreciation they lent an old diary to the author which they had bought for 6d in a 'junk shop' in the main road of Ludlow.

This diary was the beginning of intensive study of the Red Kite, being, as it was, an account covering parts of the years 1801 to 1803 by a certain Thomas Henbane whose signature, in the same handwriting and ink, was in the front of the diary. Although at first glance the booklet did not seem exceptional, its contents proved it to be a most valuable item for research. Reading the whole diary was a somewhat difficult task, mainly due to the very faded ink — in some places nearly completely gone — and also to the extraordinary formation of some of the letter characters and, quite often, phonetic spelling, mainly Welsh words.

The diary is an account of the business dealings of Henbane, mostly in standing timber. His main concern was the supply of what he terms 'long building poles'. These were the type of scaffolding in use prior to the modern metal scaffolding tubes. Henbane delivered these to all parts of Great Britain, but mostly to South Wales and Southern England. Cut timber was also supplied, presumably for burning. It is evident that Thomas Henbane was a person whose work gained him an explicit knowledge of wild-life of Central Wales. Unknowingly, his writings give such an intimate account of birds of prey and animals — such as the pine marten — that he stands on a par with any contemporary naturalist. Henbane became an expert on account of his particular trade. Although he had dealings in timber, both standing and felled, he enhanced his income by selling various birds of prey, both dead and alive, to those who wished to buy them. In almost all cases, the name of the purchaser, and the town or village, and the price paid, is written in the diary. Most of all, his accounts of where and how he obtained his specimens is of the greatest importance. The names of the parishes, inns, the proprietors or tenants, names of woods and districts are given — all, of course, for his own benefit for future trade.

To an ornithologist it represents the ground-work for studying a particular species at a period when little was put into print and the author set about the cataloguing of the itinerary. It seems that the keeping of birds of prey in aviaries, both large and small species, was a commonplace pastime, and the Red Kite was a ready saleable item. Nowhere else is there such an account of the distribution of the species and therefore an especial study was undertaken of the Kite in Central Wales. Thomas Henbane makes no less than 53 references to Kites, from whence they came and to whom they were sold. Not that he was in any way cataloguing the species, but he, no doubt, kept references for future commercial use. He sold Kites to persons as far away as Birmingham and indirectly, to customers in London and Derby.

When the time came for the two American friends to return home, the author had not quite finished the cataloguing of the diary and these kind persons actually delayed their departure for a few days in order that the author might complete this task.

CATALOGUING THE THOMAS HENBANE SITES

First of all, a complete list of places where Kites were seen, or referred to, was made. Ordnance Survey maps 1:25,000 (two and a half inches to the mile) of the entire part of Central Wales were purchased and all places marked on them. Studying these areas gave an incredible amount of information on the distribution of the Red Kite in the relevant years. Some woods had obviously existed where now none stood. Some had been planted or replanted since then. Some valleys had been flooded by the construction of the dams around the Elan Valley and elsewhere. For twelve years the author studied these remote woods. Perhaps Henbane studied these areas also — it seems he must have known the spots where Kites were most likely to be encountered. As he was taking young birds for sale, he would not waste time, presumably, on unlikely areas.

A most exciting observation unfolds — the now steadily increasing numbers of Kites during the 1970's are invading the old Henbane territories. Due to the sporting firearm bringing about the near extinction of the Red Kite, much study has been precluded until very recent years. As will be seen from the discourse on 'Protection' (Chapter 3), such attempts to help the birds did little good in building up the numbers of Kites; indeed, as soon as it was given out that the birds were to be specially protected, this put a price upon both the birds and their eggs, so that forced protection is, and always has been, detrimental to the species. Previously, Kites were shot only on account of their predations upon farmers' feathered livestock and then by the game-rearing keepers. Now skins and eggs were made a valuable item by publicity.

The come-back of the Kite was due to two things: first and foremost to the two world wars, and secondly, to a lessening of the interest in sporting estates and game-rearing. Both these incidents lowered the numbers of keepers and keepered-land considerably, allowing more breeding success. However, the greatest help of all must have been the introduction by the ornithologist Robert Beaney of Spanish Kites' eggs during the late 1950's and the 1960's. These eggs were placed into Buzzards' nests and the young Kites helped to swell the numbers of the indigenous birds. At first, Beaney brought the eggs over himself, two or three clutches per year, but later he was helped by the English ornithologist Peter Parkin and his experienced associate Jose Garcia, both based in Spain. Parkin had been carrying out experiments on birds of prey, including the introduction of eggs from one species to the nest of another, in schemes to enhance the introduction of selected birds into areas where certain species are becoming rare. His experiments could lead to the saving of threatened birds in many parts of the world.

Another source of introduction was by C. H. Gowland. In his publication Birdland, pages 365 and 366 he states he arranged for 'a large quantity' of newly laid Kite eggs to be brought over from Spain in the spring of 1927, and again in 1928 on a much larger scale. These eggs were placed in the nests of Buzzards. Over 30 young Kites were hatched during these two attempts. Gowland states that this compares favourably with the 'Kite Protection Society's' figures of only sixty-six young in nearly forty years. However, as most eggs were placed outside the 'Kite Valley' region, he assumes that a great many fell to the keepers' guns and gin-traps.

One point must be considered in relation to these imports. First, oologists find that the Spanish Kites' eggs were, on average, more inclined to be blotched and shaded, sometimes with a very pale orange-red, whereas the original British Kites had the well-known 'scratchy' markings. Many Spanish eggs lack these markings. Definite areas of birds do tend to create 'types' in eggs, even in relatively small areas. For instance, although the Stone Curlew is extant only in small parts of Southern England, it may be divided into special areas, say East Anglia, South Downs, Salisbury Plain and Berkshire. Eggs from these areas, when kept separate, show a definite distinction — even in such a small combined area. Kites, therefore, as far away as Great Britain and Spain would be more likely to locality diagnostics. It might also be a point in connection with this that Kites are now more frequently nesting in small birch trees as opposed to the habitual oaks. The latter, however, remain the most selected tree. Most of these oaks are quite small in size, although this is due mainly to the habitat and the type of weather associated with the wild open country of central Wales.

Kites did become so low in numbers that it must be assumed that Robert Beaney's and Gowland's Spanish introductions were necessary to give boost to the Kite population, even if slight changes do occur in, say, egg markings and possible choice of nesting sites. The birds themselves are probably indistinguishable, even under very close analysis. It is these self-undertaken Kite-boosting experiments which caused Milvus milvus to increase quite well during the late 1960's and the early 1970's, and not at all to any 'protection' methods. Reports show that such advertised protection creates a market for the birds in question and their eggs, to the detriment of the species. Left alone, the birds do a far better job.

It is not possible to say what percentage of nesting Kites were known to Thomas Henbane. Certes, by spreading out all the Ordnance Survey maps and placing them together, it can be seen that most of central Wales has been covered. As nearly all the natural nesting sites have been covered, it is possible that the Henbane Kites represent a high percentage of the population in his time. If this be taken at 80 percent — a purely arbitrary figure — then the number of nesting pairs would be about 65 for the area covered — an intelligent guess at saturation point. The Henbane sites were all typical (see Chapter 4 — 'Finding Kites' nests'). As Kites nested from Cardiff to North Wales a maximum figure of 112 pairs is put forward for the date of pre-1800. This takes into account that the majority of the birds would be confined to the central area, being more in keeping with their requirements for feeding and nesting purposes. Kites, as with most species of birds, hold territories, but the area of these seems to vary tremendously. The Golden Eagle is far less tolerant to invaders of its territory than the Kite. Although two pairs of Milvi might nest within a few hundred yards of each other, their territories would spread out in opposite directions from the nests and it is probable that they do not overlap to any extent, at least near the actual nesting site. However, two pairs of birds may often be seen quartering the same area with no noticeable animosity. This occurs, of course, more often during the winter months, but may be witnessed when the birds have eggs or young.

Wolley's few accounts of Kites' nests — but extremely welcome in a dearth of such knowledge — proves that nesting Kites bred throughout middle and south-east England. Impossible as it is to guess the exact numbers and distribution, Wolley's accounts do prove that Milvus milvus was indigenous to the flatter farming lands. From all sources: Henbane's Welsh records, Booth's from North Britain and Wolley's Central and South-Eastern England and taking into account the descriptions of Kite-hawking from the old sporting books, we can make a reliable guess that the Red Kite was widespread, even if not common.

CHAPTER 2

(i) IDENTIFICATION

The Kite itself is not likely to be confused with any other bird except by very inexperienced persons, when Buzzard and Osprey might give the wishful thinker problems. Once seen, a Kite will, then, always be known. By visiting the suggested advantagous points given in Appendix A, it should not be long before a Kite arrives in view. Magnificent at close quarters, the true flying capabilities of the birds far exceed any other raptor when seen high in the air in the teeth of a gale. The birds revel in their mastery over the elements, twisting and turning, diving and climbing with the utmost ease. No wonder indeed that Kite-hawking was the most princely of all forms of falconry. When pursued by a falcon, even the humble rook shows flying abilities far exceeding its normal progress through the air. It is true, then, that such a master as the Red Kite could take on and often beat in pure tactics that magnificent terror of the air, the Peregrine Falcon, whose stoops reach a speed of over a hundred miles an hour and are capable of decapitating a partridge in flight. In fact, one Peregrine alone would never succeed in bringing down a Kite. A cast of falcons (two in number) was absolutely necessary if a flight should succeed. Flights at Kite often ranged over five miles from start to finish — and that not in a straight line but with battle tactics ranging the whole time. 'Tis well worth a drive of two hundred miles to watch, say, three or four Kites in adverse weather conditions. Possibly the best place in all Wales to observe this is from the hill above the church at Cwmystwyth. Seldom are Kites not to be seen here and the surrounding mountains cause thermals and venturi in which the birds delight.

The Red Kite makes an excellent anemometer. With practice, the strength of the wind may be gauged by the amount of wing expanded. In a gale-force wind, the bird's wings are nearly closed, with the primaries lying parallel to the body. The less wind, the more sail is required and, in still air, the wings are fully expanded.

(ii) NEST

The Red Kite's nest is very similar to those of the Buzzard and the Carrion Crow in that they are constructed in the same situations in a tree. Buzzard, perhaps, require a more substantial base, as their nests are usually heavier and larger. Kites require a lesser fork to hold their weight and, in this respect, are more like a crow's abode. The main tree selected by the Kite is the oak, but lately more and more birds are selecting birch trees and even beech. A few have nested in larch and once, at least, in an ivy-covered tree. Scots pine are infrequent in Wales but E.T. Booth writes that the Scottish Kites in his time habitually used these trees. He also states that in areas where trees were very scarce the Kites nested on rocks. The height of the nest from the ground varies between ten feet and fifty feet. Only once, at fifty-eight feet, was that height exceeded. The average of thirty-eight nests was twenty-eight feet. Reports of nests at sixty feet are unlikely and were probably Raven or Carrion Crow. Ravens, in particular, can be found at enormous heights and one, near Llanfihangel-nant-Melan, at the top of a mighty larch, seemed to disappear into the very heavens and was seventy feet above the ground. Each bird has its favourite height and if one nest is deserted, the second will usually be found at a similar height to the first.

Nest usage varies between the individual birds. There can be three types: the first uses the same nest for many years in succession. The second may have one, or even two, alternate sites to be occupied in different years. The third type is the bird which seems to make a new nest every year. The second category, with two nests used alternately, is more frequent. Next is the bird which uses its original eyrie continuously.

Dan Theophilus, Allt-yr-Erw Eyrie No. 34. His wood held a nesting kite in the early 1900's G. Bowes

Allt-yr-Erw Eyrie No. 34. Kite's nest was in top right corner G. Bowes

Graham Bowes, with son Andrew above the Towy Valley and Dinas (Ystradffin) behind *G. Bowes*

Allt-yr-Erw Eyrie No. 34, barbed-wire placed there in early 1900's G. Bowes

Upper Towy Coed Allt-y-Berau Eyrie No. 27, nest in far wood — *G. Bowes*

Colonel H Morrey Salmon 'The Father of Bird Photography' *G. Bowes*

Llyn Brianne Eyrie No. 24, 1988, nest in wood on left — *D. Hughes*

Bronmwyn Eyrie No. 75 *G. Bowes*

Dinas Eyrie No. 26 — *D. Hughes*

Ty-Mawr Eyrie No. 8, The Author at Ty-Mawr, 4 May 1987, nest of 1978 still in good condition after nine years. G.W.R.D.

Ty-Mawr Eyrie No. 8, 1978 C/2 — *M.J. Dawson*

Abergwesyn Eyrie No. 1, C/2 1973, cork-screw shaped tree M.J. Dawson

Devil's Bridge Eyrie No. 15, 1985 M.J. Dawson

Rhuddallt Eyrie No. 25, 1965 — G. Bowes

Glan Dwgan Eyrie No. 13, C/2 18.4.1979 *M.J. Dawson*

Bont-Maen-Ddu Eyrie No. 63 G. Bowes

The main diagnostic feature of the nest is that it is practically flat. It has a far shallower cup than either Buzzard or Carrion Crow. This may become even flatter after incubation has started, mainly due to the frightful conglomeration brought to the nest during this time. Main debris is definitely sheep's wool. Some nests are so covered that they resemble a dead sheep in the tree and draw attention to its whereabouts. Large balls of mud seem very frequent and it is doubtful if they just fall from the bird's feet. They are most certainly gathered intentionally. This is not unreasonable as many birds cement their nests together with gathered mud — corvids in particular. Paper can be seen at times and, these days, pieces of plastic bags as used on farmlands. Rag is also found at times in the construction of the nest, but it is not nearly so prevalent as some authors would suggest. Most farm substances are now packed in the universal plastic, and the once common rotting sacks, with pieces adhering to barbed-wire fences are a thing of the past. The Kites, too, have, of necessity, up-dated their building materials.

The best time to look for Kites' nests is mid-March. At this time, the birds are prospecting their territories and fly around their chosen areas. There is also no problem with disturbing the birds at a nest with eggs. Then again in July, when the young birds are flying around the nesting wood, is the second best time. In fact, although the young are not visited very often by the parent birds, they can be vociferous and are easily seen.

The best and most certain feature of a Kite's nest is the very flat top, a fact which precludes any other ownership. Secondly, all Kites' nests have exceedingly large amounts of wool in their construction. Buzzards never use wool and the Carrion Crow uses that substance only for the lining. Third in identification value is the exceedingly filthy state of the nest, particularly when the young have hatched. The nests of most predators may have a certain amount of flesh and bones around the rim, but nothing compares with the Kite's abode. One very important and completely diagnostic feature is that Kites' nests nearly always face north-east. Very few indeed vary many points of the compass — about 5 percent may vary as far as east-by-north. This can help in differentiating the nest from that of a Buzzard, which tends to face more to the south. Neither species, however, faces west. By stating that the nest faces north-east, this signifies that the nest-tree is situated on a slope which falls away to the north-east. One might think that this habit would be a hindrance in inclement weather, but no birds of prey are much affected by the cold. They are, however, very susceptible to heat and the young in particular would suffer from a nest in the full heat of the summer sun. Not all birds of prey do construct nests facing the coolest compass point. Golden eagles may be found in all directions, with perhaps, very few due south. The majority of Kite nests are found between the nine hundred feet and one thousand feet contours. This would probably constitute about seventy-five percent of all nests and any suitable woods at that height a.s.l. should be inspected first.

(iii) EGGS

The indigenous Kites' eggs were nearly always certain when compared with the Buzzard's — the only other egg which can possibly cause confusion. Kites' eggs are more pointed at the small end and have small blotches of a rusty-red, not quite like any Buzzard. The Kite also has the 'scratchy' markings which are diagnostic. Buzzard's eggs may be found with a similar type of marking but are infrequent. Under analysis, the shell of the Kite differs somewhat but would not show up to the naked eye to any extent. Presumably ultra-high magnification of the egg-white would be conclusive, but old, blown eggs are rather out-of-date for this procedure! Fourteen clutches of old Kites' eggs, taken prior to the introduction of Spanish eggs, all conform to the above principles. A series of twenty-three clutches of eggs examined since the introduction shows a complete overlap with the usual type of Buzzard's egg. A series of Spanish eggs shows that many do not have the noticeably pointed end and that colouration is similar to Buzzards' eggs. The pale red-orange smears of the Buzzard are found in the Spanish Kites' eggs.

It is no longer possible to state for certain to which of the two species a clutch pertains and it is imperative to see the bird at the nest. Also, as a Buzzard and Kite often nest within forty yards of each other, a bird overhead is no indication to the ownership of a nest. As Spanish eggs may be confused with the Buzzards', the nest now remains a more certain guide. In most cases it can hardly be confused. Nevertheless, the bird should be seen on the nest for positive identification.

As far as is known, Kites have not been subjected to 'farming'; that is, they have not been induced to lay inordinately long clutches, as, for instance, have Sparrowhawks. The average clutch for the Kite is two to three eggs. Four, it seems, have appeared more often than have been recorded. Any such c/4 occurring is immediately taken as a grand prize and nothing is recorded. Also, c/3 could be taken before they have a chance to evolve into four. The only traceable c/4's from Wales are clutches taken in 1908 (by Major W. H. Milburn on 16th June; 1917 and 1921 (these by the same oologist); 1933, 1946; and in 1968 the author found a c/4 in the Gwydre Wood, Myddfai on 19th April, and five days incubated. No doubt others are about and not publicised. Clutches of four are frequent, but not common, in Spain; out of three hundred nests about one in eighteen were fours, according to Paul Parkin. Where Thomas Henbane mentioned the actual numbers from a nest, it seemed fours were about the same incidence as the Spanish birds, or more frequent.

Infertility in the older Welsh birds has been given as a reason for the many eggs of Kite which never hatch. Such high numbers of infertile eggs have never been recorded elsewhere in the country. In the author's opinion, it is the product of the forced protection of rare birds which caused such a large number to be proclaimed 'infertile'. Out of, say, four hundred clutches laid in just over twenty years, the number 'officially' claimed as infertile is just on fifty percent. No other bird has such a diabolical infertility rate. Presumably Kites in Spain would be, on average, subjected to the same cause of non-germination, but infertile clutches are few and far between. One reason given for such a cause in the Welsh birds has been that the stock became so low that many birds became sterile. However, since the Spanish introductions, the young of successful Welsh birds would not have this infirmity, so such a conclusion would not be applicable for the last ten years, at least. But Kites' eggs are still being declared 'infertile' at the same high percentage. Based on the author's finding of 'pricked' Buzzards' eggs in Kite's nests, it is estimated that at least seventy-five percent of the declared infertile eggs might possibly have hatched, if left alone. 'Pricking' eggs is an old gamekeeper and water-bailiff trick, which causes the birds to keep incubating on eggs which will never hatch. Should the eggs merely be taken or destroyed, the birds would lay again immediately. Also, putting pricked eggs into the Kite's nest cannot have been one of the measures considered necessary by the so-called 'kite committees'. Even if the Kite's eggs were removed for safety and put into Buzzard's nests, there would be no logic in pricking the Buzzards' eggs put under the Kite. Surely it would be better for the latter to hatch some young, and so remain in their vicinity. Indeed, better still would be to leave the Kite's nest empty, when the bird would lay another clutch and thereby help to swell the Kite population by two hundred percent.

Another reason given for infertility in Kites' eggs is the use of pesticides. So much has been written on this subject that it need not be considered here. The fact is, the pesticide problem has been over for some years now. It seems probable that some persons have been issued with permits, allowing them uncontrolled access to Kites' nests. Should such a person be seen at any eyrie, they are merely 'examining' it. The issue of no such licences would be in the interest of the birds.

(iv) BREEDING AGE

Most large birds do not nest in the year following their hatching and this applies to Kites. Even the larger gulls are three years old before they breed. Golden eagles are approximately five when they first nest, a fact which can be ascertained by the extent of white in their retrices. Kites are probably two or three years old before pairing. The introduction of the Beaney eggs would, therefore, not take effect for some time. Eggs introduced in 1959 would not produce breeding birds until circa 1961 or 1962, and the early 1960 eggs would start to show an improvement in Kite numbers about 1965, which, in fact, shows up in the proclaimed numbers of Kites in the relevant years.

(v) INCUBATION

Every text book on this subject gives the incubation as from the first egg. As no person has really studied Kites until now in detail, it is obvious that one author has merely copied another down through the years. Indeed this must be so. A writer on the birds of a whole country cannot be an expert on every detail of every species, and such persons only have an interest in birds as a whole. Just because one notices that a bird is sitting on its eggs, it is not necessarily incubating. There is a vast difference between the two. Young Kites in a nest hatch within a few hours of each other and an interval between the first and the last in, say, a clutch of three eggs is, at the most, twelve hours. Had the parent bird incubated from the first egg, then the young would hatch at two or three day intervals. A case in which this happens is in some of the owl family, where chicks of vastly different ages can be found — often to the detriment of the younger ones.

Nearly all birds — most passerines decidedly — sit on their eggs at night. Taking the case of the Long-tailed Tit, with a clutch of twelve or so eggs, should the birds be incubating while sitting on their eggs, which they do every night, then the first-hatched chick would be flown ere the last had left the shell. In sitting on eggs, as opposed to incubating them, the bird just lowers itself down onto its eggs. When actually incubating the feathers on the underparts of the birds are first expanded outwards. Then the bird lowers itself on to the eggs and the feathers then contract round them, enveloping them to give the necessary warmth. The Kite, therefore, merely covers its eggs until the last in the clutch has been laid, when incubation proper begins.

The time between the laying of each egg in the Kite varies. The second egg is laid two and a half days after the first. That is, if one egg is laid in the morning of one day, then the second is laid late afternoon two days later. Should a third egg be laid, this will occur three days after the second. It has not been possible to calculate the time lag between third and fourth eggs due to their infrequency. The seventy-two hour space between the first and second egg is not an unusual occurrence and some waders lay at one and a half day intervals. Not all birds have twenty-four hour laying periods, or multiples thereof.

The Red Kite does, however, cover its eggs until the whole clutch is ready to be incubated and the reason may be to protect its belongings from predators, both avian and mammalian — the latter including squirrels and pine martens. The female Kite undertakes most of the incubation and leaves the nest on an average four times a day. Sometimes, but not always, the male will take over and this usually occurs during inclement weather. On a warm, sunny day, both birds fly off and the male then leads his mate to carrion. At other times, food is brought to the nest, sometimes right on to it but mostly to a tree nearby. Incubation lasts twenty-eight days for each egg. When both birds leave the nest together, which will be about 16.00 hours on a fine day, this is the only time during incubation when the contents of the nest may be inspected legally and ethically, as recommended by Bruce Campbell and I. J. Ferguson-Lees in their book 'A Field Guide to Birds' Nests'. This is, in fact, the finest book ever written for those who seek nests of all types. (see pp. 139 and 145: Marsh Harrier and Montague's Harrier respectively, both Schedule A birds).

(vi) DESERTION

Here we have one of the major fallacies written in connection with the Kite. It is not possible to say who started this statement, but it has certainly been carried on down through the years by all writers on Kites. It is doubtful, however, if any of these actually studied Kites intensively over many years. Once a statement has been made, there are many persons who take that as proven and consequently reiterate such mistakes. Indeed, the Chaffinch is far more likely to desert a nest, with or without contents, than is Milvus. Merely let a Chaffinch see you looking at it while it is constructing a nest and it is more than likely to leave it and begin elsewhere.

Kite are well used to the constant traffic of farmer and labourer past its nest. A well-known fact is that Kites build in full view of a busy farmstead and sometimes, even before the whole village. The kindly farmer near to the Strada Florida Kite at Crofftau told me that the female Kite sat continuously, while a number of forestry workers planted young conifers to within a few feet of its nest tree. A few Kite have nests which command a view of a busy road, with no consequential detriment to their incubation. The great John Walpole-Bond, one of the finest craftsmen in field ornithology ever, said that a Kite will not desert if the first egg of a clutch is taken or lost by predation.

The author has watched people, presumably armed with licences allowing them uncontrolled deliberate disturbances, approaching Kites' nests and deliberately putting the birds off the nest. They then have 'examined' the nest and, having thoroughly disturbed the bird, move on to the next. Unfortunately, no one has explained to the Kites that these persons may do such things. To show exactly how much tolerance a Kite does possess, one farmer told the author that a Kite on his land was subjected to visits almost every day by 'watchers' who made sure that the eggs were still there. Despite this continual disturbance, the eggs hatched, but the young died at about one week old. No detrimental effect upon the eggs occurred, but the young Kites, left uncovered, died from exposure. However, continual 'checking', maybe two or three times a day, by self-appointed 'protectionists' could be a great hazard.

Desertion, then, among Kites is no more prevalent than in most species and the bird is, in fact, much more tolerant to disturbance than many others.

(vii) FOOD

The main part of the Red Kite's diet consists of carrion, mostly dead sheep and lambs. The latter seem to have a higher mortality rate than in other parts of Britain due, mainly, to the structure of the land and its poor soil. Adverse weather conditions on the higher land also account for sheep losses. The main feeding areas are the open mountain tops — Frydd areas — and steep-sided valleys. The harder the winter, the more carrion there usually is available. Most raptors fare well during hard weather, as their prey is more easily taken. Perhaps the exception is the Kestrel, which feeds on small mammals and these become scarce under thick snow. Ravens do particularly well in a hard winter, and show this by nesting earlier than after a mild winter. As with the Short-eared Owl, breeding habits, including numbers of eggs to a clutch, adjust themselves to the food supply.

Various small mammals and birds may be taken by the Kite, but reports of full-grown dead rabbits would be beyond the capabilities of a Milvus getting it into the air. It requires a female Goshawk to take rabbits well and these accipiters have a grip four times greater than that of a Kite. Few male Gosses take rabbits well and neither sex of gentilis can lift a rabbit off the ground or even drag it along with ease. E.T. Booth states that the main food supply of the then Scottish Kites was grouse. He also mentions squirrels, rabbits (young or defunct adult) and the young of Curlew, duck and pigeon.

(viii) AGE OF KITES

Kites have been kept in captivity for up to twenty years, but this does not necessarily give an accurate guess of longevity in the wild state. In natural surroundings, birds have to run the gauntlet of many hazards, such as weather and availability of food. Goshawks kept by falconers often live ten years and cases of fifteen and twenty years are known. Golden eagles live for forty years in captivity.

The one way of recording the age of birds is by ringing them or marking them in a visible way. The ringing method only works on the birds being 'procured', usually by the shotgun. Very, very few birds are ever found dead, although odd cases do arise, perhaps when they have been poisoned and are becoming weaker when they may collapse in an area frequented by humans. The only other method is that of the experienced oologist. Through their eggs, a female bird may be traced over the years by her own distinctive egg-markings. Most birds are stable as regards their own breeding area, even though they may travel many hundreds of miles out of the breeding season — or even during that time of year, such as Shearwaters. So far, not much work has been carried out in this way on the Kite. Certain 'privileged' people have been issued with licences deliberately to disturb the Kite, but, in the opinion of the author, they have not been experienced in field-craft.

Taking into account cases of other birds of prey which have been closely studied, the usual age of the Kite, supposing it be spared a charge of shot, is around twelve years. It is possible that a few more years could be the average, but until the right people are issued with licences and are able to study the Kite even more closely, the above must be regarded as the average.

It is extraordinary how persons not suitable for certain jobs are give certain tasks. A scheme for introducing White-tailed Eagles was once carried out. The person in charge had the birds enclosed in wire netting surrounds until they were fully summed (all feathers hard-penned) and then the wire was removed. Immediately, all birds took wing and most disappeared in a short time. It was 'supposed' that the eagles would take time to be efficient on the wing. Now, any experienced falconer could have explained that the eagles should have been 'hacked'. They should have had no enclosing wire and have been left on a hack loft while still unable to fly. Being fed at 6 am and 6 pm *within a few seconds,* the eagles would have learned to fly gradually, and would have returned to the hack loft (a high mound of grass with vertical sides) dead on time, for food. They would have become used to their environment and so would be more likely to remain in the area. The author took the trouble to write and explain this method — used for centuries — but, of course, received not even a reply. We read of another 'expert' who said he could tell the age of Hen Harriers by the colour of their eyes. Falconers have been able to estimate the age of all light-coloured-eyed birds for centuries! The dark-eyed falcons, of course, cannot be estimated in this way, but may be told, although not so easily, by their feet. It would be interesting to note the colour-change in Kites' eyes over the years. Being a light-eyed bird, these organs should be diagnostic. Falconers know other methods of gauging the age of various raptors and these might be applied to the Kite.

CHAPTER 3

THE PROTECTION MOVEMENT:

An impartial look into this matter, in relation to the Kite in particular, is appropriate. When the Kite population became so low during the early 1900's, some very well-meaning persons took it upon themselves to try to protect the remaining birds in Wales. Not only did they give their time to the scheme, but also helped in a monetary sense. However, literature shows that their efforts really did little good and after twenty years, the number of Kites remained almost the same. Reports of actual numbers of birds vary with different writers. In 1903, for instance, John Walpole-Bond gave four pairs and one odd bird. Oliver G. Pike states three pairs only for the same year. The Handbook, the ornithologists main tome, gives ten to twenty pairs up to 1920. While one author gives eight nests for 1910, the Handbook states ten. The lower numbers given by various societies and writers are certainly not so accurate as the estimate of ten to twenty by the Handbook.

It is incredible how some persons — styling themselves as 'experts' — are unable to find nests, even of the easily-found Kite. One such person, giving evidence in a court, said he found the Abergwesyn Kite in 1971, but 'was unable to find the nest in 1972'. The fact that the Kite moved to its alternative site, about eight hundred yards away, completely baffled this 'expert'. Lack of field-craft is, of course, the reason. Such persons see, but do not observe what they have seen. Therefore the extremely low numbers of Kites for the years 1900 to 1920 are probably based on observations by not really experienced persons. In 1962, the number of nests was still given as fifteen. The author agrees with Witherby in that the average throughout these years was around this figure. The numbers of Kite, therefore, did not improve at all until the introduction of Spanish eggs by Beaney and Gowland,

and these required another few years before beneficial effects were seen, say around the mid 1960's.

It is most unfortunate that the protection schemes immediately put a price upon Kites and their eggs. Until this time, the main antagonist was the hill-farmer, who only shot birds which attacked his feathered-livestock. The game-rearing fraternity, of course, were feuding with anything with a hooked beak at all times. Now, of course, the skins and eggs became a valuable acquisition. In order to carry out the protection of the Kite, persons had to be engaged and these then held the problem in their hands. Some, at least, cashed in on the new racket. One writer states that a party motored down from London and took Kites' eggs one dark night, this with no time to search for the eyries or the birds. Obviously, they were taken to the site and were no doubt on their way back to the Metropolis within an hour or so.

Most damage has been done by the ridiculous 'prices' put on birds and their eggs by protection societies. One person caught taking a Peregrine falcon said that he had read that the bird was worth two thousand pounds! In actual fact, he was not able even to give the bird away. Another person took two clutches of Osprey in one year, a c/2 and c/4, but was not able to present them gratis to anyone. It is probable that not one person in this country is now a 'dealer' in eggs. In other words, eggs taken by others are of no value. Certain persons, or classes of persons, are blamed by protectionists for the low numbers of some birds — or even birds which are no rarer than has been normal for centuries. They usually fall into two classes: falconers and egg-collectors.

HEIGHT OF FALCONRY:

From 1850 to 1939 numbers of Peregrines were taken by falconers for their sport and yet all traditional eyries were always occupied. Some nests in particular were in demand, as the birds which came from these were deemed excellent flyers. One such eyrie was that on Lundy — and yet, year after year, these birds bred in the same situation with no loss of numbers, despite nearly all, if not quite all, the young being taken. Guy Mannering (B.O.A. Bulletin No. 52, November 25, 1936) in a discourse on the birds of South-East Kent, states, of the Peregrine Falcon: "six to eight pairs breed every year on the cliffs; their numbers have varied little in the last forty years. All young were taken every year for falconry up to the war". And so it has been all over the British Isles. No matter how many Peregrines were taken, their numbers have never been affected.

HEIGHT OF EGG-COLLECTING:

From around 1840 to 1939 nearly every boy, at least in the country, collected eggs at some time. There were many hundreds, even thousands, of mature collectors, even businesses thriving on the purchase and sale of eggs. And yet the numbers of birds were at a pinnacle. Collecting diminishes no species whatsoever. Indeed, one pair of Magpies devours more eggs during a breeding season than all collectors could possibly do in a number of years. And there are many thousands of pairs of Magpies — and other predators, such as Jays and Carrion Crows, and the mammalian predators, such as mice and squirrels.

HEIGHT OF THE SPORTING SHOT-GUN:

Here we have the one cause, up to 1939, which did become a menace to birds of prey. The great John Walpole-Bond, who possibly knew more about ornithology than all the present-day protectionists put together, knew that the taking of eggs did not harm the numbers of birds, but he was livid when a bird was killed, whether common or rare. A bird may lay again many times, even in a season, and all through its life, but to shoot a bird meant killing the actual source of reproduction.

The great classic example is that of the Scottish Ospreys. Had Dunbar and St. John collected only eggs, the species would have survived. They, however, shot out all the old birds, screwed the necks of the young and took the eggs. Dunbar wrote: "I think we have done for the Ospreys in Scotland!" The shot-gun was the only reason for the decline of the Kite population in Great Britain.

DECLINE OF CERTAIN BIRDS SINCE 1945:

There are three main reasons for such decline.

1. DESTRUCTION OF HABITAT:

This is the greatest cause for alarm and a subject upon which all ornithologists agree. Should land disappear under money-making schemes, either good or bad, then the birds would be annihilated, in which case the bird-watcher would be without his chosen hobby.

2. PESTICIDES:

A short-lived but extremely dangerous — to man as well as to birds — cause of bird decimation and particularly to the raptors. Much has been written upon this matter, and all ornithologist are aware of the results. However, action was taken just in time, and such birds as the Sparrow-hawk and the Peregrine are now nearly fully recovered. The latter species has now only to occupy its more southerly eyries to be back to full strength.

3. ECOLOGICAL FACTORS:

This is a matter which man cannot attempt to adjust. Due, possibly to climatic changes, certain species of birds are restricting their breeding range. Some, whose area extended only just into Great Britain are becoming rarer (Wryneck/Red-backed Shrike, etc.). However, just as these certain species disappear, so do others colonise this country and enable ornithologists to see birds which their fore-fathers were unable to do (Collared Dove, Savi's and Cetti's Warblers, etc.).

To return to the Kite, we find that, in reality, only the shot-gun was responsible for its low numbers.

CHAPTER 4

FINDING KITES' NESTS:

It is a fascinating pastime looking for birds' nests. A most remarkable book appeared in 1972 entitled 'A Field Guide to Birds' Nests.' Its authors are two of the most experienced ornithologists in this country, Bruce Campbell and James Ferguson-Lees. The whole volume is given to the finding of nests of all birds breeding in Great Britain and, in this author's opinion, has more information in this respect than all other ornithological books ever printed. It is a 'must' on the shelves of all persons so interested. Of course, such a book, encompassing around two hundred species, cannot go into such complete details as can one book on one bird and dealing with just nest-finding. Nevertheless, it shows that a large number of persons are extremely interested in such a craft — and a highly specialised field-craft it can be. The finding of Kites' nests may be divided into three sections: Time, Place and Methods, and they will be taken in that order.

TIME:

First must be pointed out that looking for the nest of this bird, while it has eggs or unflown young, may constitute an infringement of the law, as one, especially as he becomes more acute at field-craft, comcs upon a nest inadvertently and thereby disturbs the nesting Kite. The best time of all is the last two weeks of March. The trees have, as yet, not broken out into leaf, which makes nests more visible and the birds have not started to lay. The odd Kite does lay in the first week of April, although the main body is not 'down' until the second or third week of that month.

Kites will, at this time, say the 14th to the 31st March, be in their 'home' areas and will be flying in and above the wood in which they will be constructing the eyrie, or, perhaps, relining an old nest for use. They may be watched in the air, with continual glides and stoops into the home wood. This is also the only time when any antagonism may be witnessed between territory-holding Kites. The better the weather, the better will be your chances of seeing the birds, as they often stay perched during inclement times.

The second time to seek Kites is during the beginning of July or perhaps a week earlier. Now the young are on the wing, but frequently fly back to the nest, particularly at night. They remain in the same wood — if, indeed, the nest is in such a place. If in an outlying tree or a thin belt, the young Kites repair to the nearest deciduous wood. Thick conifer plantations are never occupied at all, neither by young or old birds. When the young are flying, there is no risk of falling foul of the law.

PLACE:
Kites have nested from the North to the South of Wales. At the present time, it would not be very profitable to seek the birds in their old haunts such as Devon, Lincolnshire and Speyside. Even the extreme northern and southern parts of Wales are, as yet, mainly without Kites, although it is hoped that they will eventually spread into these areas. At the present time, the area to be considered is that encompassed by the Ordnance Survey points SN6080 in the North-West to SO2080 in the North-East, and then down to SN6020 in the South-West to SO2020 in the South-East. Of course, a few Kites may be found outside this area, perhaps more to the North-West, but over forty pairs now breed in the above area. This number, by the way, is getting on for half saturation point, in that the number of suitable territories is about twice that figure.

METHODS:
The method of finding a Kite's nest may be divided into two categories, Modern and Scientific. We shall consider the first

of these two and state, at the beginning, that the modern way is purely to let the self-styled protectors show you exactly where the nests of these birds are situated.It must also be pointed out that these methods are not a very satisfying way of carrying out a pastime which may be both exciting and require great field-craft. Nevertheless, such Kite abodes may be found by these methods and are, unfortunately, becoming more and more in use by those who are not really ornithologists, but are the product of the false propaganda exuded from some protectionist. That such persons already are acquainted with these methods makes the publication of them of no extra value. They are given here for completeness.

(1) THE BARBED-WIRE METHOD: Probably the simplest of all and involving no skill whatsoever. Merely walk through the most likely woods in Kite country and look for nests of a large size in the trees. Should the trunk below a nest be festooned with a mass (rather mess) of barbed-wire then that is a Kite's recent abode. No other species is treated in this way in this area. Trip-wires surrounding a tree, either at ground level or up to five feet high, also indicate a Kite's nest tree.

(2) WATCHER METHOD: This is a product of the modern age. One walks along the roads in Kite country. It should not be long ere a person descends upon you and orders you off in no mistaken terms, as you are disturbing 'his' highly protected birds. You are now within a hundred yards or so of a Kite's nest. A notable point is that this person *never* owns the land and is a self-appointed protectionist who may eventually cause the Kites to desert their nests.

(3) LICENCEE METHOD: Should you see a vehicle of the cross-country type in Kite country, it is as well to observe its meanderings. It is possible that the occupants, if they be not the local farmers, are such as are issued with a licence, deliberately to disturb the birds, and these

powers they use to the full. They march boldly up to a Kite's nest, 'inspect' it and having thoroughly disturbed the Kite in all lawfulness, proceed to the next nest. One may find half a dozen nests in a few hours by this method.

(4) CAR METHOD: One or more cars parked in an out-of-the-way place, possibly with no amenity or scenic value, can usually be attributed to so-called 'watchers'. They park as near as possible to a Kite's nest and finding it gives no trouble.

(5) ADVERTISEMENT METHOD: One of the simplest means of all to find any rare bird's nest is to read the advertisements in a bird protection magazine. They state the rare birds in their area and more information is available on arrival. One such advertisement stated 'Two pairs of Golden Eagle breed on our 5,000-acre estate'. As there were only two suitable crags, one could walk straight to the nests.

So much for an assortment of 'modern' methods. Many more may be devised; not a very satisfactory way and not appealing to the student of ornithology. Such methods may be left to those who have read the false propaganda that birds or eggs are worth many thousands of pounds and are trying to cash in on a racket which does not exist.

SCIENTIFIC METHOD OF FINDING KITES' NESTS: The true ornithologist will prefer the method based on field-craft. Shall we say that he has never visited Kite country, even that he has never travelled to the Principality at all. Yet, when he has studied the maps, together with collation of all he knows, then he may drive to that part of the country with a certainty of finding eighty percent of the Kites' nests there. After studying the following facts, he will be able to do so.

There is a most interesting observation by the experienced ornithologist Don Humphrey (J.S. Bulletin No. 117) on the finding of a Greenshank's nest by picking it out on the map before even visiting the area. As he points out, another ornithologist had already made a statement that a nest was found by sensible map reading. Of course, not all birds' nests may be pin-pointed on a map with sure success. The Kite lends itself to such scientific analysis.

Therefore, let all known facts be collated herewith:

AREA: Already described under the heading 'Place'. To the student of caliology, he would do well to remain within this area, from Devil's Bridge in the north-west to the northern slopes of the Brecon Beacons in the south: from the Black Mountains in the east to Lampeter in the west. Here procure the maps of the relevant area as suggested in Appendix B. One cannot even start without the Ordnance Survey maps and these are second only to binoculars in importance to an ornithologist. The One-Inch maps are more useful than the newer 1:50,000 in that they differentiate between the coniferous and deciduous woodlands. As far as the Kite is concerned this is an absolute necessity.

The requirements of nest-finding in the Kite may now be listed and then each one taken and analysed separately.

(a) Nests face north-east
(b) Usually situated on a steep slope.
(c) Preference for oak woods.
(d) Height a.s.l. usually between 800ft and 1000ft contours.
(e) Possibly overlooking a farmstead.
(f) Large areas of Frydd in immediate vicinity.
(g) No large or extensive forestry plantations completely enveloping the wood.
(h) No major road within 800 yards at least.

After having located the best places with reference to the above, further considerations may be exacted when actually visiting the places.

These may be sub-divided as follows:

(i) Oaks not too closely planted.
(ii) Height of main trees.
(iii) No Carrion Crows nesting in woods.
(iv) Height of nests in wood 15ft to 35ft.
(v) Gauging height of nests from ground.
(vi) Analysis of nesting sites.

Take the first section:

(a) NESTS FACE NORTH-EAST: About fifty percent of all species of birds tend to place their nests in a certain compass direction. Usually, of course, one finds that small passerines take advantage of the sunnier side of a location, but this is by no means always the case. Nest boxes should not be placed in the full heat of the sun, preference of selection being towards the east. Birds of prey are not unduly affected by the cold and, in fact, the opposite is the case. Heat is very detrimental to young hawks and even to mature birds, which often take to high-flying to cool themselves. The Kite is particularly affected by the sun. Of course, some vary a few points and these are to the east usually, with a few to the north. Neither Buzzard nor Kite ever face in a westerly direction. Carrion crows differ more than Milvus. By stating that nests face in a north-easterly direction, this means that it is built on a slope which falls away in a n/e direction, that is, where there is the least amount of mid-day sun. Should a nest or its contents come to grief, the bird will then repair to the next available and suitable wood facing the same compass direction. Seldom does it use the same wood for its alternative, although the Abergwesyn Kite did so on one occasion, using a twenty foot tree on the lower edge and about a hundred and twenty yards to the south-east (the nest still facing the correct direction on the slope). It is remarkable how some persons are unable to find a nest. As mentioned already, one self-styled 'expert' said that he was unable to find the Abergwesyn Kite in a particular year. The fact that the bird used its alternative site, about eight hundred yards away in the next wood, completely baffled this person.

The Ordnance Survey One-Inch maps are the best for use in picking out suitable woods, in that they differentiate between coniferous and deciduous woodlands by symbols. Larches, being a deciduous conifer, are given the coniferous markings. The latest O.S. 1:50,000 maps show no symbols for the two types of woodlands and are therefore not so useful in this respect. They are, however, more easily read in that they are reduced only to one and a quarter inches to the mile. The two and a half inch maps are really the ultimate for ornithological use. They often pin-point even a single tree, marked 'oak' or 'elm'. However, they may not be, in the First Series, completely up-to-date as regards the latest afforestation areas and also a few of the new reservoirs. This may be overcome by marking any extra woodlands, etc., upon them by reference to the new 1:50,000 maps. The author also uses a green crayon-type pencil to colour the woodlands so that they stand out. The new Second Series maps are already distinguished so and are up-to-date.

An interesting point in Kite sites is that they are changing somewhat since the introduction of eggs by Beaney and Gowland. The inherent instinct in birds from Spanish eggs is manifesting itself in such aspects as choice of nest site. In place of oaks, birches and beech are being used. Also, one or two much taller trees are being used; a Spanish trait. Welsh Kites were never very fond of trees over thirty feet and Booth mentions the Scottish Kites nested between fifteen and eighteen feet.

Perhaps the best method to show how Milvi do select these north-eastern slopes may be seen by studying the O.S. map 1:50,000 number 160. Take a look at the area to the north and west of Brecon. This is indigenous and truly excellent Kite country. Mainly open frydd area for feeding, the land is split into mountain ranges by streams, all running from north-west to south-east. In these valleys are small woods of oak. Some are on the northern slopes, *facing* south-west, which may by ignored. The woods opposite are on the south-western sides of the streams and are all Kite-potential woods in that they fall away to the north-east. Few of these have never held a Kite's nest in the last hundred years. Most of this part of Wales must have been in such a wild state before the afforestation of the land. Kite have to make the best of this and, by applying the north-east rule to the other parts of the terrain, most nests of this species will be found. One's thoughts must be adjusted to areas, swamped under conifers and flooded for reservoirs, and select the next-best areas, just as the Kites themselves have to do. The Handbook states: In 1933 to 1937 five or six pairs of Kite in one area but only four nests located. The object of this book is to enable an ornithologist to find them all and in a short space of time, by field-craft.

(b) USUALLY SITUATED ON A STEEP SLOPE: By and large, as the mountains drop into the valleys, the sides become steeper as they get lower. It is on these slopes that the oak woods hang and in such woods the Kites prefer to nest. There are, of course, exceptions to prove the rule, such as the Kite at Crofftau, near Strata Florida Abbey. No doubt, as Kites become more common they will be forced to accept sites which may be described as not ideal. A hundred years or so ago, there would have been innumerable suitable sites, but, over the last century, woods have been felled, coniferous plantations have nullied the frydd areas and ideally situated positions have become scarcer. However, in order to pick out, on a map, the most likely situations, one should go for the steeper woodlands.

(c) PREFERENCE FOR OAK WOODS: Over ninety percent of Kites still prefer the indigenous woods of oak. As remarked before, one may find Milvus nesting in birch, beech

Kite C/2 Tŷ-Mawr Eyrie No. 8, 1978 — *W. Webster*

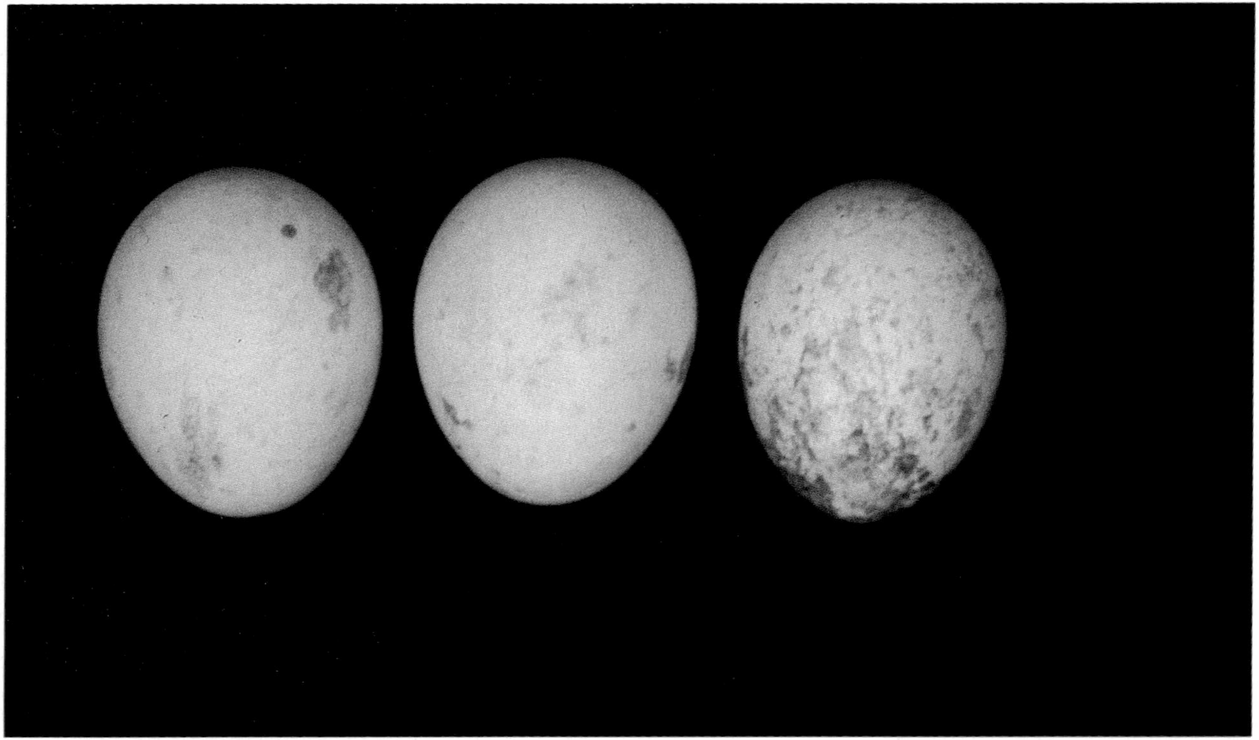

Kite C/3 N. Gilroy (1907), Bryn Crwn Eyrie No. 119

W. Webster

Rhyd-y-Groes Eyrie No. 28 — *G. Bowes*

Eyrie No. 85 *G. Bowes*

Llannerch-y-Cawr Eyrie No. 6 *Young Kite after leaving the nest* *M.J. Dawson*

Cwm Chwefru Eyrie No. 72 *M.J. Dawson*

Kite *G. Bowes*

Dinas Hill 1938 Eyrie no. 26 *Ron Nichols*

From one Kite King to another, Dinas Eyrie No. 26. The photograph was taken in this wood G. Bowes

Llannerch-y-Cawr Eyrie No. 6 *D. Hughes*

Kite C/3 Brechfa Eyrie No. 43, 20.4.1979, M.J. Dawson *W. Webster*

Kites, see two — G. Bowes

and larch and, once, in a scots pine. An unusual site is at Llannerch-y-cawr SN 902613, where the Kite's nest is in a huge larch, and extraordinarily, out on a limb. Mixtures of oak and beech are quite commonly seen and, in these, the Kite may choose either species of tree. On one occasion an ivy-covered tree was selected. Oak woods, however, still remain the best source of success. The trees themselves should not be too closely planted or, if they are, there should be ample open spaces therein.

(d) HEIGHT ABOVE SEA LEVEL USUALLY BETWEEN THE 800ft and 1000ft CONTOURS: This may be useful in deciding which woods of a group are the more likely to contain nesting Kites. Rarely above the 1000ft mark, they are occasionally found at a lower level. This applies to the country here being considered. Naturally, in the eighteenth century and earlier, when the Kite was found in Lincolnshire, Kent and so on, the height a.s.l. would be in accordance with those counties.

(e) POSSIBLY OVERLOOKING A FARMSTEAD: Reports of Kites being extremely shy and therefore keeping well away from human habitation may be ignored completely. Over half the Milvus nests are within sight of a farmstead or farm buildings. Some of these are now vacated holdings but were, in the past, inhabited. Reports say that even a whole village was in sight of some Kites' nests. These raptors, when feeding on the ground, do not usually take wing until an observer is within a hundred yards or so. Here again may the myth, that Kites are easily caused to desert while incubating, be dismissed.

(f) LARGE AREA OF FRYDD IN IMMEDIATE VICINITY: This is an ecological necessity for the Kite, in that its main feeding habits are concerned with carrion. It therefore requires a sufficient area on which to find such food. One thousand five hundred acres would be the minimum for each pair of Kites. Where there are no areas of frydd, there will be few nesting Kites in Wales.

(g) NO LARGE OR EXTENSIVE FORESTRY PLANTATIONS NEAR: This needs qualifying to some extent. As in (f), plantations must not be completely surrounding a wood which, in all other respects, might seem suitable. If, however, a plantation is to one side, say, one quarter of the circumference, then this would still leave enough feeding area of frydd to satisfy the pair of Kites.

(h) NO MAJOR ROAD WITHIN 800 YARDS: So far the author has not found a nest within this distance. The incessant noise, especially in the summer months, would interfere with the quietude of the birds. Although Kite are not disturbed by the presence of man, even continual, they do require tranquility in their domain.

The above should be sufficient to enable the caliologist to ascertain, from the maps, the exact woods in which the Kites will nest. Once there, one may make a closer inspection to ascertain if the wood is, in fact, as suitable as it seems to be by map reading. Although oak woods may be picked out cartographically, symbols do not distinguish between thinly planted and closely planted woods. One may now consider the sub-divisions:

(i) OAKS NOT TOO CLOSELY PLANTED: Both sparsely and thickly planted woods are used by Kites. In widely spaced trees, say twenty paces between, the Kite may nest in any part of the wood. If they are thickly planted, then the nest will be nearer the edge or sometimes in a boundary tree, such as Crofftau, or on the perimeter of a clearing. It is possible that the bird will nest in the thicker parts of the wood where the tops of the trees, particularly the nest tree, are open at the top, affording good flying exit and arrival.

(ii) HEIGHT OF MAIN TREES: These will be between twenty and forty-five feet in total height. Naturally, woods with larger trees may well contain a Kite's abode, but we are here dealing with the average bird. When in a birch tree, the total heights may be even less.

(iii) NO CARRION CROWS NESTING NEAR: While the Kite and Buzzard are at times within forty yards of each other, there will not be a nesting Carrion Crow within three hundred yards of either species. Therefore, if one comes across a Crow's nest with eggs or young, one may confidently move off for the above distance before expecting to come across an inhabited Kite's abode, or one that will eventually contain eggs. Corvus corone drives off most large birds from its territory.

It is as well to note here that a Kite's nest with eggs or young may only be inspected while *both* adult birds are away, in order that they be not disturbed. See Campbell & Ferguson-Lees: A Field Guide to Birds' Nests, pages 139 (Marsh Harrier) and 145 (Montague's Harrier).

(iv) HEIGHT OF NESTS IN WOOD 15ft to 35ft: Although Kites have nested as low as 10ft from the ground (1978 near Rhayader), the main body tends to start at around twenty feet. The majority do not nest above 40ft or so, indeed, the average is around 28ft. E.T. Booth stated that the Scottish nests were 15ft to 18ft from the ground. He also remarked that if there were no suitable trees in the vicinity, then the Kites nested on rocks. Henbane mentions the *rocks* at Rhydoldog in connection with a Kite's nest but there may be some ambiguity here, as he writes of a Buzzard at the same time. Kites are often around this area and nested in woods long since felled. It could have been possible to locate a nest here. We are reminded of the fact that Booth had seen these birds nesting on rocks in Scotland. The area around Rhydoldog is suitable and the rocks themselves have innumerable small trees clinging to their sides. Not an impossibility; Kites were reported nesting in this area (Treheslog) within the last sixty years. It would be fascinating to have at least one rock-nesting Kite in the Principality. The Rhydoldog rocks face north-east. Any nests over forty feet may be Carrion Crows' abodes or even Buzzards'. Here again, one does find the exception, so that all 'tall' nests cannot be ruled out.

(v) GAUGING HEIGHT OF TREES AND NESTS FROM GROUND: By keeping in practice and climbing trees out of season, one may drop lengths of string therefrom and then measure the string afterwards. There are various ways of height-measuring and these have been explained in the Appendix D.

(vi) ANALYSIS OF NESTING SITES: Usually in steep hanging oak wood, facing north-east and more often in the lower half of the wood. Nests 20ft to 35ft up in tree. Usually against trunk but sometimes out on a limb. Medium sized nest, looking very bedraggled and non-used, *always* flat-topped, nearly always covered in sheep's wool. No carrion crows nesting within 300 yards. Frydd areas adjacent; often farmstead within view of nest. Usually between 800ft and 1000ft contours. No major road within half a mile. Wood not surrounded by forestry plantations.

CHAPTER 5

The Kite Man of Pontrhydfendigaid

by Graham Bowes

Some nine years have passed since I first set out to fulfil my childhood dream, that of finding the Kite in Wales. I had read of its mysterious existance in a hidden, remote valley, depicted by a photograph of its nest in a tall tree, which appeared to be shrouded in mist. It was in one of those monthly magazines that carried all sorts of natural data. I only wish I had kept it. Somehow I imagined that if I found the valley, I would find every Kite in Wales. Now that I know it was the Towy valley, I realise I was not far wrong. For at one time, that place was thought to be the last refuge of this mysterious creature. Perhaps I knew as a lad one day I would venture forth and find the hidden spectacle in the hills of Wales. What I could never have dreamed of knowing was that after nine years, my efforts would culminate in the almost total knowledge of the creature, its secrets both past and present and the personalities and organisations that surround it. Even beyond my wildest dreams, I could never have imagined that I, Graham Lionel Bowes, would be known to the dalesmen and kinfolk of Pontrhydfendigaid as 'The Kite Man'. I can only imagine that my deep-rooted enthusiasm and thirst for knowledge kindled their bestowing upon me that truly treasured accolade. Perhaps it is only they who know I have walked through the snow and the rain, just to catch a glimpse of the Kite, to find its secret habitats in those rain-sodden hills.

I would like to record here my thanks to all the people who in their own particular way have helped me to know the Kite. Firstly to my dear departed friend, Colonel Morrey

Salmon, one of the early pioneers of Kite preservation, founder of the Kite Committee in 1949, 'father of bird photography', to quote Eric Hosking, 'the doyen of Welsh birdwatchers'. His inspiration to me through our conversations and correspondence kept me alight on the trail of the Kite. Just two weeks before he died I told him I had seen Kites from Llanddeusant in the south to Machynlleth in the north. I feel somehow he knew then I had won the day. His enthusiasm and love of the Kite remained to his dying day. Even when I first knew him some five years before his death he told me how he had driven to Rhandirmwyn from his home in Cardiff, had seen a Kite, but couldn't drive any further on, the snow was too heavy. He was nearly ninety then!

To all the farmers and people in the Kite area of Wales who have imparted their knowledge to me. In particular to Dan Theophilus of Rhandirmwyn who remembers all the old pioneers of Kite history, Salter, Edmondes-Owen, Meade-Waldo and of course Morrey Salmon himself. They all climbed his hillside at Allt-yr-Erw where Kites nested when at their lowest ebb. It was he who told me that, when down to three pairs, the places where they nested. His wood was one of these.The barbed-wire protection against egg-thieves is still there, around the tree. Oh, Dan! for your memories I yearn.

Not least my thanks must go to the author of this book, whose prowess and skill in tracing the haunts of the Kite both past and present have led me to the very depths of its existance. My sincere thanks to you all, diverse though may be your attitudes, without your help I could never have climbed the mountain.

Much of my inspiration has come from the few books and publications that have recorded the Kite. From Colonel Salmon came 'The Red Kites of Wales', which depicts the protection of the species from the early part of the century. I find this very sad reading as it tells of people who appear to have devoted their whole lives to the Kite only to have died in frustration and embitterment in the knowledge that the creature remained perilously near to extinction. Miss Dorothy Raikes and the Reverend Edmondes-Owen, whose graves I have visited, were two of those highly credited by Colonel Salmon. It tells of the many hazards that faced the Kite in the early part of the century, not least the egg-collectors who at that time were a major threat to the Kite's existence. One such atrocity he records: 'the village policeman at Cilycwm tried to sell a Kite's egg to a visitor!' He records that in 1922 Edmondes-Owen considered three pairs to be the 'whole remnant' and indeed that may well have been true if the Rhandirmwyn area was their only considered habitat. But in the latter pages of his book Salmon opines that Kites were probably nesting elsewhere in the central Wales area 'unnoticed and unrecorded'. I have no doubt this was correct; one such habitat must have been in the Strata Florida area. Having spoken with Evan Edwards of Berthgoed, he has told me that both he and his father remembered Kites being present in that area. I feel very sorry for the early pioneers of Kite protection, but happy that Salmon lived through to this time of the creature's prosperity.

Another valued publication given to me by Peter Davis (Nature Conservancy) is his complete and detailed study 'The ecology and conservation of the Red Kite in Wales'. This is an unemotional and clinical analysis of the Kite's habits and status both past and present. I suppose it is not for me to say that such recording is or is not of use to the Kite's prosperity, though I tend to think that left to its own capabilities the Kite can do very nicely. It will swarm again if left reasonably unmolested. Perhaps the greatest value of such a publication is that it might provide useful evidence for those legislating protection measures. However, no legislation can have any effect on perhaps one of the Kite's worst enemies, the weather.

Dr. J. H. Salter was one of the early pioneers of Kite protection and his publication 'The Kite in South Wales' defines its habitat from about 1870 to 1920. He mentions the 'remote woods' above Strata Florida and that the Kite lingered there. As I have already said I believe it never left there. Other places where he knew the creature lived were Upper Chapel, Merthyr Cynog, Llanfihangel Nant-bran, Rhayader, Devil's Bridge and Sennybridge, all of which I have visited in the last nine years. It gave me enormous pleasure to meet one Phillip Davis at Merthyr Cynog, who remembers Kites being present there in the early part of the century. He also vividly described many of the characters such as gamekeepers who Salter had mentioned in his book. One such man, Thomas Griffiths, he said was a tall man and when seated, would run his ginger beard through his hands like a piece of rope! As with Salmon, Salter tells of the many human inflicted hazzards that have hindered the Kite's survival. As I have mentioned earlier Salter had often visited the nesting site at Allt-yr-Erw, Rhandirmwyn, home of my dear friend, Dan Theophilus.

Without doubt, more than to any other publication, I feel I owe much of my recent progress and inspiration to the author's first edition of 'The Red Kite'. In this he reveals a wide range of habitats of the Kite both past and present, many of which I may never have found of my own accord. He has also kindly furnished me with another publication 'The Natural History of the Kite', by James Fisher. This depicts the secrecy of the Kite's habitat in Wales, protection history and its population in Britain in previous centuries. Perhaps of particular interest are the details of how eggs were imported from Spain and planted in Welsh Buzzards' nests in an effort to increase the British Kite population. Whilst the success or failure of this project remains unknown, there seems ample evidence that indeed it occured.

I feel I should mention the sheer magnificence of this creature, the superb shape and form of the 'three-winged bird', for with its bending, twisting, swallow-like tail that is exactly what it is. It has been written that no other bird of prey uses its tail as does the Kite. The long narrow wings which help it to drift for hours only a few feet from the ground. 'Like a blown leaf' is an apt description of its movement. Never more so have I seen this phenomenon when in a bitter cold winter in 1981 I watched a Kite gliding slowly over the back gardens of the little town of Tregaron. I remember standing in the car park there somewhat elevated from the town, with home fires giving off the smell of burning wood and watching the Kite drift slowly on. A magic moment indeed! Then the magical colouration of its plumage. A superb combination of rust, black, white and various shades of brown. Perhaps for me the white patches under the ends of the wings, sharply contrasted with the dark wing tips, enhance most the beauty of its colour. From above, the 'horseshoe' arrangement of its feathers and the contrasting colour of its rusty-red tail once again give it a most striking appearance. Needless to say I could watch them for hours and indeed I have done so. Something which I think spoils their appearance and which I find distasteful is the 'tagging' of the birds. The tags used are clipped to the 'elbows' of the wings, are extremely evident and spoil photographs. They also give the creatures an 'air of captivation'. I am not convinced that they serve a useful purpose apart from some sort of identification.

A lot of attention has been drawn to the Kite through the media of egg-collection and collectors. Although I assume the eggs are collected much in the same way as perhaps are coins, cards, etc., some of the collectors would appear to be also deeply interested in wildlife and in particular the Kite. I remember when Colonel Salmon was driving me back to Cardiff station one afternoon, we were talking about the highly reputed John Walpole-Bond. He said "He was an egg-snatcher, you know". That apparently was true, but at the same time Bond had given Colonel Salmon his book 'Bird Life in Wild Wales' and had annotated the columns with the exact locations of nesting-sites that Bond had depicted in the ordinary text with code letters. I feel sure my dear friend must have found this of enormous benefit in locating his 'favourite bird'. That very same book is now in the hands of Peter Davis and no doubt he too has found it 'useful'. So although egg-collecting cannot be condoned, and I cannot understand why it is done, (why not collect chicken eggs?), the followers of the cult may well have served some indirect purpose.

I feel sure that as well as to the Kite itself I am drawn to its haunt by the continually changing weather of central Wales and therefore the ever changing appearance of the landscape. Whilst most of the time the hills are covered by mist and rain, when the sun does shine through, it illuminates the land with an almost fluorescent light. And if the Kite is seen in that same light its superb colour is highlighted to a phenomenal degree. I remember one late winter afternoon after it had rained all day I was driving from Cwm Ystwyth through to Pont-rhyd-y-groes and stopped on the road junction that leads down to the Ystwyth bridge. There above me was a 'circus' of about twenty Kites caught in the brilliant light that shone from the setting sun from beneath heavy cloud. The colour spectacle that filled the sky could only be described as breathtaking, even to an impartial observer. I have often passed by that point again in the hope of recaptivating that spectacle, but alas this has not been my good fortune. I have, however, seen many similar spectacles; for instance above Blaen Caron at Tregaron one October afternoon a mixture of about twenty or thirty Kites and Buzzards in a similar lighting condition.

Without doubt the centre place of attraction in Kite lore must be Rhandirmwyn and the surrounding district. No other place in general has inspired so much attention. Indeed, apart from my thoughts of Strata Florida, it was at Rhandirmwyn where Kites made their last stand. Among the sacred woods in which they made their refuge were Pen-y-rhiw-iar, Allt-rhyd-y-groes, Ystrad-ffin, Rhyddallt, Allt-yr-Erw, Aberbranddu, Trausnant, Dalarwen, Troed-rhiw-bylchau,

Gellihernin and other smaller glades.

Its existence in these places is surrounded by an almost fairytale aura, for it has to be said this is one of the most beautiful places on earth. The fact that when the Kite was at its lowest ebb this area was wild and unkeepered and it must have made it even more secluded and spectacular. Sometimes, even now, especially in winter, I can feel the past is still there. I remember well on a cold, dreary December afternoon I was standing below Allt-Maesymeddygon where the Towy meets the Doethie and above me a Kite drifted along the sky-line. Only a silhouette, but that which represents the past history of the creature. For here in this place it survived when extinct elsewhere in Britain. As the light fades, the silence is only broken by the gently flowing river. There I stood where the Kite pioneers of the past had held their lonely vigil. Perhaps while current day strife continues, there is yet hope that this may still be the last refuge of peace for men and the Kite. For me though, the most beloved haunt must be Strada Florida, in particular the wooded hillside at Berthgoed where in October, 1979 befell my good fortune of seeing for the first time the Kites of Wales! I shall never forget on a late October afternoon, the display afforded to me by six of these angels of the sky. The colour, the form and the magical manoeuvrability were there for me alone to see. Thank you, dear Lord! Of all the sightings I have had of Kites, there are needless to say some that stand out in my memory more than others. Apart from my original sighting at Berthgoed, an excellent exhibition in itself, I recall other moments of watching the Kites. Another late October afternoon when I stood with Idris Jones of Tregaron at the slaughter-house there, to see eight Kites at equal distance between each other just above the hillside. Following a dreadful winter week in 1982 with massive snow-falls and sub-zero temperatures, when Peter Davis thought perhaps they had all perished, I related to him that once again from the hillside at Berthgoed I had seen six rise to the sky.

I remember they followed me all the way down to Tyn-y-cwm and remained at a great height above the farm; as if to say, for you Kite man, we've made it! Perhaps for me they'll always make it. The time when I was talking to Evan Edwards of Berthgoed, he on his horse and in driving rain, a Kite battled against the wind over a nearby field. Perhaps in much the same way one would have done a hundred years ago. For this is the ancient home of the Kite! More so I believe than Rhandirmwyn. My first winter visit was in January 1980, when the harsh cold and snowy conditions had driven the Kites close to Tregaron. Just below the old slaughter-house they dived repeatedly at corvids in the yard there. The sighting of the pair at Pennal near Machynlleth told me they had stretched their boundary. In October 1987 I saw a Kite fly over the wood at Allt-yr-Erw; a vision of the past? Perhaps my loneliest time with the Kite was in February, 1986 with the temperature way below zero, I walked from Rhandirmwyn through the Cothi valley past Aberbranddu and then to Cwrt-y-Cadno. One was gliding along the top of the crag above Aberbranddu and then as I climbed the hill at Cwrt-y-Cadno which leads to Farmers I saw another Kite through the over-hanging trees. Don't seem to remember seeing any bird boffins that day! I have watched the Kite in places abroad; on the Spanish side of the Pyrenees, the morning a little train took me from Canfranc to Jaca, a Kite followed the train down the hill! And in Minorca where Salmon had read there are more Kites (Red) than anywhere else on earth; I can well believe that. The bird's the same, but it's not Wales!

It has been said that it is not possible to set up a nest viewing sight as has been done for the Osprey and the Peregrine and I would agree. But what is possible is a public hide set up at Tregaron rubbish tip. If erected in a strategic fashion this would afford hours of enjoyment for those wishing to see the Kites at close range. I feel sure that once familiar with such a structure, the Kites would accept it as they do other permanent buildings. I ask this to be considered by all parties concerned.

So what has driven me through pouring rain and blinding snow over these last nine years? Perhaps the same yearning for the delights of natural history as I had for the Great Crested Newt with their warty skins and yellow bellies; huge toads that lurked under stones; the Pike with its formidable array of teeth; the yellow collars of the Grass snakes I once kept — one ended up on the step of the local grocers shop! Sand-Martins diving in and out of their nesting holes in a sand bank. The Adder I coaxed into a brush-case. Little Owls perched on telegraph poles along the Pilgrim's Way. Lizards on stones and posts on the North Downs. Caddis fly larvae covered in tiny shells and river fragments. The stench of the Devil's Stink-horn fungus. The super white collar of the Ring-Dove, its flimsy nest often given away by its constant cooing. And the colour in a frog's eye and the sheen in a starling's feathers. What is left? Perhaps the Lammergeier in the Pyrenees. But so sufficient has been my experience with the Kite in Wales I need wander no further. There seems little reason why the Kite should not regain its former numbers, my philosophy is such:-

The more it flourishes,
The less attention it will draw,
And the less attention it draws
the more it will flourish.

My final words must be those of thanks to the many diverse personalities who have helped me to know the Kite. Thank you Morrey Salmon, Mike Dawson, Peter Davis (thanks for the Kite feathers), Dan Theophilus, Evan Edwards, John Pentwyn, Ivor Williams and all the farmers and Welsh people I have met along the way. But perhaps most of all to the people of Pontrhydfendigaid who have so graciously left me the most treasured epitaph I could have ever wished for, The Kite Man. I thank you all,

Graham Lionel Bowes.

(It is to people with such a high interest in the Red Kite that we must thank for their continued existence in Great Britain. No one is more excited than Graham Bowes in watching Kites. The Kites were at a very low ebb, and remained so, until many eggs were brought over from Spain by C.H. Gowland and Peter Parkin. From that day, Kites began to increase. Gowland was an egger, but was also the main help in the Kites' return.

I remember walking into the Black Lion in Pontrhydfendigaid with Graham Bowes. As the heads of the locals turned, a unanimous cry exclaimed: Ah! The Kite Man! No one has studied the Red Kite more than Graham.

Mike Dawson.)

CHAPTER 6

Kite Clutches and Eggs

by John Hemmings

The average size of 58 British-taken Kites' eggs measured by Jourdain was 56.97mm x 45.09 mm, with the maximum being 64mm x 46.6mm and 60.3mm x 49mm. The minimum sizes being 52.6 x 42.9mm and 54.9 x 42mm. Jourdain gives no shell weights. H.E. Dresser gave egg measurements from his collection in inches and these vary from 2.10 x 1.75" to 2.34" x 1.80". He also gave no shell weights but the German collector Dr. Rey is more helpful and from his collection of Kites' eggs from the Continent he gives the average size of thirty eggs as 56.7mm x 44.48mm. The average weight is given as 5.37 grms.

Henry Seebohm, in 1896 notes that Kites' eggs vary from 2.4" to 2.1" on their long axis but seldom less than 1.75" on their short axis. Here again, no shell weights from him.

Schönwetter, in 1967, gives sizes from 52 to 63 x 40 to 49mm and the weight of an unblown egg as 63 grms. The heaviest unblown egg recorded was 71 grms.

Makatsch in 1974 gives an average of 48 German collected eggs as 57.35mm x 45.13mm and shell weights vary from 6.15grms to 4.34grms, an average weight of 5.23grms.

A Welsh clutch of Kites' eggs is as follows: C/4 Towy Valley, Wales. June 1908. Collected by Major W.H. Milburn. Sold at Stevens' Sale Rooms 11th January 1922. See also K.L. Skinner's publication Oologists' Exchange & Mart, March 1922. This was a late infertile clutch. Identified by the Rev. F.C.R. Jourdain as a Kite in 1922. The eggs were sparsely marked with 'tick' and 'scratch' markings. Egg 1: 55.3mm x 43.9mm. Weight 4.37 grms. Egg 2: 54.1mm x 44.4mm. Weight 5.23grms. Egg 3: 55.3 x 41.7mm. Weight 4.85grms. Egg 4: 53.6mm x 42.4mm. Weight 5.18grms.

Another Welsh clutch of Kite, a c/3, taken in May 1902. Locality: Cardiganshire. Collected by C. Jefferies. Recorded as being sold by Watkins & Doncaster in 1945 to Capt. A Pearman of Purley. Egg sizes are: Egg 1: 59.1mm x 45.9mm. Weight 5.76grms. Egg 2: 60.0mm x 45.6mm. Weight 5.09grms. Egg 3: 59.9mm x 45.8mm. Weight 5.53grms. This is a handsome clutch blotched and smudged with orange-brown, with a few scratch markings.

Kite c/3. May 1904. Locality Brandenburg, Germany. Collected by R. Tanacre. From the P.F. Bunyard collection and sold at Stevens' Sale Rooms in 1937. Egg 1: 60.0mm x 46.4mm. Weight 6.65grms. Egg 2: 59.4mm x 45.5mm. Weight 5.25grms. Egg 3: 60.0mm x 47.0mm. Weight 6.35grms. A beautiful clutch marked and blotched with orange-brown, some violet undermarks and some scratch markings.

C/3. May 1896. Locality Sweden. Collected by Helge Lilliestierna. From the P.F. Bunyard collection and sold at Stevens' Sale Rooms in 1938. Egg 1: 56.1mm x 44.3mm. Weight 4.67grms. Egg 2: 55.1 x 42.2mm. Weight 4.30grms. Egg 3: 59.0mm x 43.1mm. Weight 4.23grms. A somewhat uninteresting clutch with a few tick and scratch marks. A thin-shelled clutch with two of the eggs below the minimum recorded weight of 4.34grms by Dr. Makatsch.

C/3. March 1887. Locality Grenada, Spain. Collected by a local man. From the P.F. Bunyard collection and sold at Stevens' Sale Rooms in 1937. Egg 1: 55.0mm x 43.9mm. Weight 5.26grms. Egg 2: 55.3mm x 42.1mm. Weight 5.22grms. Egg 3: 53.8mm x 45.0mm. Weight 5.94grms. A good looking clutch, blotched with dark brown and some fine peppered marks. Also, one egg is heavily marked on the pointed end; some scratch markings are present.

C/3. April 1900. Locality Mark Brandenburg, Germany. Collected by E.F.M. Eastman. Sold at Glendinning's Sale Room in 1947. Egg 1: 57.2mm x 44.7mm. Weight 5.51grms. Egg 2: 54.8mm x 43.5mm. Weight 5.49grms. Egg 3: 55.0

x 44.3mm. Weight 5.36grms. A fine looking clutch. Heavy blotches of orange-brown. No scratch markings present. One egg blotched at pointed end.

C/3. May 1904. Locality Pomerania, Germany. Collected by Kricheldorf. From the P.F. Bunyard Collection. Sold at Stevens' Sale Rooms in 1937. Egg 1: 53.4 x 44.2mm. Weight 4.71grms. Egg 2: 52.3mm x 43.3mm. Weight 4.91grms. Egg 3: 51.6mm x 43.3mm. Weight 5.00grms. A fine looking clutch, blotched and spotted with orange-brown markings. Again, one egg blotched at pointed end. No scratch markings present.

C/3. May 1890. Locality central Wales. Collected by C.P., a local man. Sold at Watkins & Doncaster's about 1950. Egg 1: 55.8mm x 42.3mm. Egg 2: 54.5mm x 43.0mm. Weight 4.82grms. Egg 3: 56.0 x 43.2mm. Weight 4.61grms. An interesting clutch marked with dark brown. Many scratch markings and again one egg blotched with dark brown at the pointed end. This set is somewhat thin-shelled and below average weight. The sizes are above the minimum recorded.

C/3. May 1910. Locality Mark Brandenburg, Germany. Collected by a local man for Ludlam. From the collection of C.B. Horsbrugh. Sold by C.H. Gowland in 1952. Egg 1: 57.0mm x 42.5mm. Weight 4.65grms. Egg 2: 49.8mm x 42.2mm. Weight 4.01grms. Egg 3: 51.8mm x 43.0mm. Weight 3.96grms. A handsome clutch with heavy blotches in brown. Some light lilac under marking. Pale lilac scratch markings also visible. Again, one egg is blotched on the pointed end. Slightly under average weight.

C/2. April 1921. Locality Southern Spain. Collected by Souis Castel. From the Capt. A. Pearman collection. Egg 1: 55.1mm x 42.8mm. Weight 3.81grms. Egg 2: 55.7mm x 42.76mm. Weight 3.79grms. A small clutch with thin shells. Below average weight. Marked with mid-brown spots and small blotches. Some scratch markings present and some violet under markings.

C/2. April 1923. Locality Laguna della Janda, Southern

Spain. Collector: Commander Vaughan for Harry Kirke-Swan. Sold at Watkins & Doncaster in 1950. Egg 1: 59.4mm x 44.6mm. Weight 5.39grms. Egg 2: 58.3mm x 45.1mm. Weight 5.35grms. Fine looking clutch spotted with brown. Some tick markings present with mauve under marks. No scratch markings.

C/2. May 1916. Locality Central Wales. Collector was G. Tickner. From the collection of L.R.W. Loyd. Egg 1: 54.5mm x 41.1mm. Weight 4.94grms. Egg 2: 58.5mm x 43.1mm. Weight 4.80grms. An interesting old Welsh clutch, blotched with dark brown. Many scratch and tick markings present. One egg nicely marked at the pointed end with scratch markings. Again, slightly below average shell weight.

C/2. April 1901. Locality North Wales. Collector Colonel R.H. Rattray. From the collection of G.G. Wood. Egg 1: 58.0mm x 42.6mm. Weight 4.90grms. Egg 2: 59.2mm x 43.2mm. Weight 5.38grms. An interesting clutch. Small blotches in mid-brown with fine specks and tick marks. Here again, one egg blotched with heavy markings at the pointed end.

Looking through our study collection at a series of Red Kite clutches the distinctive shape of the Kites' egg becomes very apparent when compared to a Buzzards' egg. The Kites' eggs are slightly larger and usually more pointed. But very rarely, they do overlap and pointed Buzzard's eggs do, at times, turn up. Many Kite's eggs have the distinctive scratch markings which seldom occur on Buzzards' eggs. These scratch markings are also found on Black Kites' eggs which are, of course, smaller eggs and could not be confused with the egg of the Red Kite. The Rough-legged Buzzard also, at times, shows scratch and tick marks but usually is a much rounder egg. In 1927 the late C.H. Gowland, a dealer in eggs, imported from Spain twenty-one Red Kite eggs and with the help of some fellow ornithologists, placed them in Buzzards' nests in Wales in the Towy Valley area. Some of these eggs were infertile. Some hatched and at least two birds were reared by Buzzards. In 1928 he made another attempt and imported fifty-three eggs and again made use of Buzzards' nests. A much greater success was had this time with thirty-one birds being reared. It should be understood that much of the Red Kite recovery in Wales was due to the efforts of an egg-collector with the help of Oologists. Since 1927, the Kite 'egg type' has changed somewhat and the continental 'egg type' is more pronounced in the present day Welsh Kites. Much of the scratch markings have disappeared, although still present in a lesser degree.

In 1914 John Walpole-Bond described the eggs of Welsh Kites as follows: The eggs are larger and much more elliptical than those of the Buzzard. A characteristic type is of a whitish or yellowish-white ground colour, finely spotted, streaked and scratched pretty evenly over the whole surface with yellowish-brown, dark red and rust colours. The inferior markings (under-markings) are usually scant and scarce, being of lilac-grey and purplish colours. Bond goes on to describe other types as skim-milk ground colour, richly blotched, smudged and freckled, chiefly at one end, with reddish-brown, chestnut and grey. A fourth type is a dirty dead green, zoned with blood-red and dark brown. He found also unmarked eggs. The scratchy marking, he goes on to say, is almost peculiar to the eggs of the Red Kite and this was certainly correct up until 1927. In the mid-1930's, the Rev. F.C.R. Jourdain describes Kites' eggs thus: Colour white and without gloss, often feebly marked with brown, but at times blotched and spotted with varying shades of reddish to purplish-brown. Dark hair lines are rather characteristic. Many Welsh eggs are pointed-ovoid in shape, but in a series from other countries this is not characteristic and resemblance to the eggs of the Buzzard is often very close.

(John Hemmings is one of this country's most expert oologists. In fact, there is no living person with a greater knowledge of the eggs of the birds of Britain. His collection, handed down from oologists of the past, is possibly the most useful to those studying eggs of birds. These specimens are available for all those persons who study oology in depth.)

APPENDIX A

Apart from seeking the nests of Kites, some ornithologists may just wish to observe the birds in the air. The following places are suggested as some of the most likely where this species occurs throughout the year:

(i) From the hill above the church at Cwmystwyth: Glog SN794744 looking south.
(ii) Craigen Ddu SN 777655.
(iii) Devils Bridge/Bwlchcrwys SN 715776.
(iv) Rhosygelynen SN 900630.
(v) Lan-wen SN 915726.
(vi) Mynydd Mallaen SN 750440.
(vii) Abergwesyn SN 852535.
(viii) Rhiwiau SN 753265.
(ix) Moelfre SO 012485.
(x) Cefn Merthyr Cynog SN 975385.

APPENDIX B

Ordnance Survey Sheets required for the area under discussion:

1-INCH SHEETS: Nos. 116, 127, 128, 140, 141. These are the best maps to use for Kite analysis as they differentiate between the deciduous and the conifer trees by symbols.

1:50,000 SHEETS: 1¼ inches to the mile. Woods coloured green but no symbols. Nos. 135, 136, 146, 147, 160. The latest edition (1988 onwards) now have woodland symbols, as the older 1-inch maps have.

1:25,000 SHEETS: 2½ inches to the mile:

CENTRAL AREA:

SN	69 79 89 99		
	68 78 88 98		
	67 77 87 97	SO	07 17
	66 76 86 96		06 16
	65 75 85 95		05 15
	64 74 84 94		04 14
	63 73 83 93		03 13
	62 72 82 92		02 12

OUTLYING AREAS:

SN	56	SH	60 70 80 90
	45 55	SJ	00
	44 54	SH	61 71 81 91
	03 13 23 33 43 53	SJ	01
	02 12 22 32 42 52	SH	72 82 92
		SJ	02
SO	08 18	ST	17
	09	SS	84 94 (Somerset)

APPENDIX C

BIBLIOGRAPHY:

Beaney R: paper, *Breeding Ecology of some Raptors in Spain*. 1975

Booth E.T.: *Rough Notes*, Volume 1. 1881-1887.

Campbell & Ferguson-Lees: *A field Guide to Birds' Nests*. Constable 1972.

Fisher J.: *Natural History of the Kite*.

Gosnell H.T.: *Science of Birds' Nesting*. Gowland 1947.

Gowland C. H.: *The Natural History of the Kite*. (From Birdland pp 365/366)

Hardy E. *Guide to the Birds of Scotland*. Constable 1978

Hardy E: *The Naturalist in Lakeland*. David & Charles 1973

Jourdain Society Bulletin: Nr. 117, September 1976

Mitchell A: *Field Guide to the Trees of Britain and Europe*. Collins 1974 and 1978.

Morrey-Salmon H: *The Kite in Wales (from Welsh Wildlife in Trust)*. North Wales Naturalists' Trust. Bangor 1970.

Parkin & Garcia: *Selective Breeding of Spanish Raptors*.1976.

Salter J. H.: *Kite in South Wales*. Paper 1928.

Walpole-Bond J: *Bird Life in Wild Wales*. Fisher-Unwin 1903.

Walpole-Bond J: *Field Studies of some Rarer British Birds*. Witherby 1914.

Witherby: *Handbook of British Birds*. 1938.

APPENDIX D

GAUGING HEIGHTS OF NESTS IN TREES: The above and also gauging the heights of the trees themselves may be accomplished by the same means, of which there are several. After some practice, one may be accurate to within a foot up to 50ft and only slightly less exact over that height. Possibly the easiest way is to get used to the height of buildings, on which there are many thousands to practice. All the iron rainwater pipes (commonly, but erroneously, termed drain-pipes) have their joints at six-feet intervals, except the shorter lengths at top and bottom which have purposely been cut. Count the number of full lengths from ground-level to gutter: about three on a normal two-storey building and add any estimated short lengths. You cannot be more than an inch or two out. Afterwards, when regarding a tree, one may visualise the number of 6-foot lengths quite easily. A small amount of practice makes one very proficient.

Another method is to stand a person against the trunk of a tree and, allowing an extra inch or two for his height, gauge the number of 6-foot intervals to the nest or top of the tree.

Should one wish to be highly technical, the following method may be used:

One measures a comfortable distance from the tree centre (i.e. half its thickness at the bole), in this case as an example, 80 feet. Then measure the angle from ground to top of tree or nest. Using the equation Tan = Opposite over Adjacent, the height is calculated. All one requires are one's Trig. books!

$$\text{Tan } 43^\circ = \frac{H}{80\text{ft}}$$

Abergwesyn Eyrie No. 1, C/2 18.4.1975 *M.J. Dawson*

Gwydre Wood Eyrie No. 40, C/3 20.4.1981 *M.J. Dawson*

Pwllpeiran Eyrie No. 18, 1988 M.J. Dawson

Abergwesyn Eyrie No. 1, young kite and egg, 20 May 1973 — M.J. Dawson

Rhyd-y-Groes Eyrie No. 28 — *M.J. Dawson*

Hafod, Cwmystwyth Eyrie No. 19 — G. Bowes

Craig Rhosan, Cilycwm Eyrie No. 31 *M.J. Dawson*

Tyn-y-cwm, Strata Florida Eyrie No. 21, March 1978 — *M.J. Dawson*

Crofftau Eyrie No. 20, 20.4.78, Nest on edge of wood — *M.J. Dawson*

Brechfa Eyrie No. 43, C/3 20.4.79 *M.J. Dawson*

Llwyn Garu Eyrie No. 22, 1987 *M.J. Dawson*

Young Kite 1981, site unknown to 'protectionists' and therefore undisturbed — G. Bowes

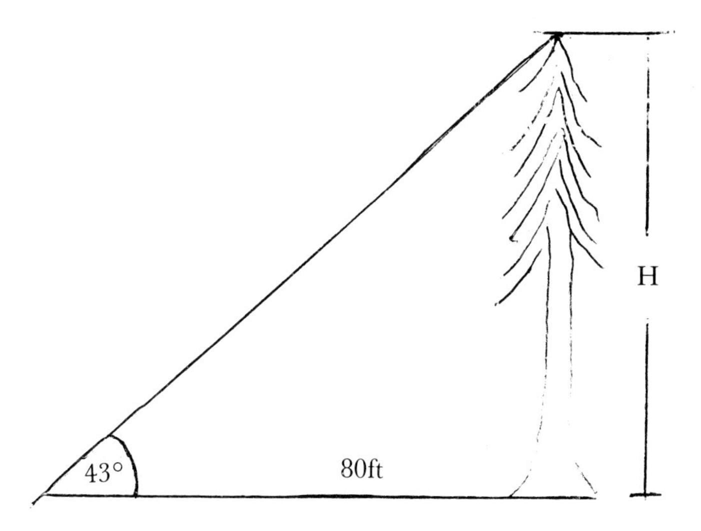

An excellent method described in the Field Guide to the Trees of Britain & Northern Europe by Allan Mitchell requires only a stick. Such a stick may be procured on the spot; old, new or bent does not matter. In any case, the book is well worth a place on the shelves of the ornithologist. Most accomplished ornithologists are also interested in many other aspects of the countryside.

Another method is to have a mirror set at 45° in a holder. Lie upon the ground and crawl forward, away from the tree, looking into the mirror. When you see the top of the tree, the height is from the base of the tree to the mirror. This all depends on the ground being level and one holding the mirror level. The above methods all depend on the ground level factor. Any up-grade or down-grade must be deducted or added.

No doubt other gadgets may be evolved to make the task simple. Estimating by six-foot intervals will be just as accurate after a small amount of practice. Any person not able to retain such a length in his mind will never become an expert in field-craft.

ADDENDUM I

Young Kites, as soon as they are hatched, show a pinkish tinge to their down covering and may thus be distinguished from young Buzzards, which have greyish-white down.

Since the practice of 'pricking' Buzzards' eggs was discovered by the author, those who have a mind to do such things are now using the 'shake' method. Once an egg has been incubated for a short while, it may be rendered infertile, or, to be more exact, the embryo may be killed, by shaking it violently for some seconds. There is then no discernible way of determining whether the egg was infertile or why the embryo died shortly after incubation started. After having been incubated, then, by a Kite, the contents become the usual watery conglomeration of any incubated-infertile egg. Such a Kite, sitting upon 'shaken' Buzzards' eggs, would eventually be claimed 'infertile'. Such eggs, removed from the nest at the end of the full incubation period are completely soiled and not in good order for preservation. They are, however, much sought after by a certain body of persons, *not* ornithologists or oologists. The laugh is, of course, upon these individual~~s, as the~~ eggs are not Kite, but 'switched' Buzzards' eggs. One person, (as yet not named, but known) has, for the last ten years or more, been switching such eggs. The author estimates the number of Kites' clutches so treated is between forty and sixty within the last fifteen years or so. The author believes, in his opinion, that this is the reason for the so-called fifty percent infertility rate in the Welsh Kite.

John Walpole-Bond writes that Buzzards are far more shy a bird than Kite and that the former nest in more secluded parts. This is yet another point to show that Milvus is not a timid bird and is not easily disturbed. Walpole-Bond, in his Field Studies of some Rarer British Birds says that Kites do not incubate from the first egg, thereby agreeing with the author of this dissertation. Bond's critical observations and keen eye could differentiate between the incubating and covering of eggs. His notes agree with this author's on-the-spot records. A paper issued about 1973 contains various assertions on the Red Kite, but most of these are completely wrong. The writers of that paper were not acute in observing details; in fact, they even ascribe to the Kite a nest with a deep cup! This has *never* been the case. Inability to differentiate between Buzzard and Kite was the reason. Walpole-Bond asserts, and rightly so, that a Milvus nest is absolutely flat before the eggs are laid, becoming, perhaps, a little more indented after incubation has started, due to debris laid around the rim. Frohawk, in his book, states that Kites' eggs and even young, were often blown from the nest in a high wind. With such a flat platform, this would admittedly be the case. This could be another point why the Kite covers its eggs until the last in the clutch is laid and incubation begins. Also in his Field Studies, Bond states that the Buzzard *never* intentionally adds wool to its nest. The Kite, of course, adds extreme amounts of wool, both before and after incubation has started.

The shell of Kites' and Buzzards' eggs differ to the expert.

The author is much indebted to the late Ron Nichols for the following information on a clutch of four Kite's eggs; this was in Stevens' sale of 10th and 11th January 1922, second day, lot 578 (part of), bought by Doncaster. Ex Major W.H. Milburn collection, taken by him on 16th June 1908 in South Wales. These details in catalogue. The late date would suggest a repeat clutch.

The list of Kite sites is not absolutely complete as some could be missed. As the Kites become more numerous they will spread to other areas and other woods. They may be found by following the advice in this book. The author would be most pleased to hear from other dedicated ornithologists studying the Red Kite on the exact nesting sites which are not given in this book.

The average size of the Kite's egg is 57 x 45 millimetres. H.A. Gilbert mentions that the egg of a Kite taken around 1905 had the curiously pointed shape typical of this species and that nine specimens he had seen were of this form. (B.O.A. Bulletin Nr. 9, January 1926).

Col. Cecil Smeed points out (B.O.A. Bulletin Nr. 16) that the Kite always uses wool in its nest as a lining and in the construction thereof, while the Buzzard never does so and uses green vegetation. He also points out that the egg of the Kite is usually pointed, while the average Buzzard's egg is more rounded. These points agree with the author and with John Walpole-Bond and with other competent writers.

A controversy in a national newspaper over the species of Kite found in London indicated that bones of the Kite found in that area and examined proved to be from the Red Kite. It was thus claimed that this was proof that this species inhabited London. The fact that Red Kites were kept by numerous people in aviaries nullifies the point. Competent ornithologists are of the opinion that it was, in fact, the Black Kite which scavanged the streets of the Metropolis.

LIST OF KITE NESTING SITES

Eyrie No:	Map	O.S.Ref	Name	Observations
1	147 SN	846530	Abergwesyn	C/2 1973. C/2 1975. Nest pushed out of tree by 'Protectionists' to stop eggers taking the eggs. Has nested further up valley and at edge of field to south of this site.
2		857538	Glangwesyn	Protectionists said they 'were unable to find this nest'. C/2 1974.
3		901512	Nant Einion	Nested here in the early 1970's.
4		967569	Brynieuau	An old site, but Kites seen there in last few years.
5	SO	020492	Blynbanedd	Nested here in 1902. J. Walpole-Bond's 'own' particular site.
6	SN	902612	Llanerch-y-Cawr	Has nested here for the last ten years or more. Nested in low Oak, then in very high Larch. 1988 in lower position again.
7		912651	Glannau	Mentioned by Henbane in the years 1801 to 1803 and is nesting here at the present time.
8		922730	Ty-maur	c/2 1978. Nest still in good condition in 1987, nine years later.
9		896753	Dernol/Tan-yr-Allt	Nested here for many years, but not recently.
10	SO	070732	Ty'n-y-berth	Nested once in late 1970's.
11	SN	820748	Ty-Mawr	In small group of scattered Oaks. C/3 in 1986.
12		752738	Caermeirch	Nested in 1987 but eggs taken.
13		706749	Glandwgan	In small open wood at SW corner of the lake. In Beech. C/2 1979.
14	SO	043433	Groes-wen	Nested here in the parish of Gwenddwr in the 1890's.
15	SN	711784	Old Shafts	One of the many sites along this valley, Cwm Rheidol. C/3 1981.

Eyrie No:	Map	O.S.Ref	Name	Observations
16		745766	**Mynach**	In bottom of steep cwm in high Oak. C/3 in 1983. One egg disappeared.
17		774746	**Experimental Farm**	Original nest in 1984 was near the cottage. Moved to Site No: 18.
18		773747	**Pwllpeiran**	Nested 1987 and 1988
19		755728	**Hafod**	In woods south of caravan site. Has nested in this area for many years.
20		747648	**Crofftau**	Nested at edge of wood in 1979. Female continued incubating while young conifers were being planted. Nest pushed from tree by conservationists.
21		771649	**Tyncwm**	Probably the oldest known site. Henbane mentioned these woods. Still used most years if birds not molested by protectionists.
22		705592	**Llwyngaru**	Nests here regularly. 1986 in small tree. In 1988 moved to a tall slender oak. Trip wires around this tree showing disturbance.
23		766517	**Nant-Irwch**	Walpole-Bond inspected the eyrie here. Visited by Arthur Whitaker on 8th April 1946 and again on 6th June 1946 when it had one young. Wood now demolished.
24	147	SN 814496	**Llyn Brianne**	On south-west shore of the east arm of the dam. C/3 1987.
25		793477	**Rhuddallt**	A very old site. Mentioned by Henbane, Ron Nichols, Walpole-Bond. Nested in 1965 and again in 1966 when the eggs were taken.
26		783468	**Dinas Hill**	Probably the most famous Kite site of all time. The double-peaked hill is a famous land mark. Photographed in 1898 by Kearton

Eyrie No:	Map	O.S.Ref	Name	Observations
				and in his book Rarer British Breeding Birds, pages 55 and 57. The Twm Shon Catti is on this hill. Seen by Henbane, Walpole-Bond, Ron Nichols and Arthur Whitaker.
27		774471	Craig Alltyberau	Another old site. Nests may be in any part of this woodland, along the river. Used in most years.
28		762482	Pen-Rhiwiar/ Rhyd-y-Groes	Continuous woodland with 27. Nested in 1987 and 1988.
29		744471	Gwenffrwd	Used many times over the last hundred years. Harrassed by Protectionists in latter years.
30		715461	Aber Branddu	Old site near the farm. Used 1986.
31		751426	Craig Rhosan	An extremely steep wood. Old site, seen by Walpole-Bond and still used in some years.
32		758432	Cwm-Y-Rhaiadr	Alternate site to 31. C/3 in 1985.
33		773435	Dinas Fawr	A frequently used site.
34		767442	Coed Allt-yr-Erw	Dan Theophilus's wood. Nested in early 1900's when barbed-wire was nailed to tree. Still attached.
35		996410	Pantycelin	Cambridge-Philips visited this nest as described in his book Birds of Breconshire. The keeper, Tom Philips saw the last nest in 1896 and birds remained until 1903.
36	SO	010472	Aber-tochen	Nested in 1985. Birds often seen in this area.
37	146 SN	639417	Nant Troyddyn	Used in a number of years.
38		646407	Penarth	An alternate site to 37.
39		695435	Pen-twyn	Used in the early 1960's.
40	160 SN	784277	Gwydre Wood	The only c/4 seen by Dawson, 19th April 1968. Incuabtion about 5 days. C/3 in 1980 and again on 19th April 1981. Nicely marked, pepper-spot eggs.

Eyrie No:	Map		O.S.Ref	Name	Observations
41			785282	**Rhyblid**	Alternate site to 40. Also in a tree nearer to farm in 1988.
42	146	SN	595609	**Bwlch**	This strip of woodland stretches for a mile and a half from the B 4342 up to Llan-faelog. Kites have nested here in parts of the wood for a number of years, especially in the early 1980's.
43			604603	**Brechfa**	In small wood just over ridge from the road opposite Brechfa farm. C/3 on 20th April 1979. A very high nest and awkward climb.
44	135	SN	726730	**Maen Arthur**	Tried to nest in 1981.
45			695883	**Moel Golomon**	A frequent site in recent years.
46			677880	**Pen Dinas**	The alternate site to 44. 1977.
47	135	SN	748966	**Cefn Coch**	Nest usually in top half of wood, which is unusual.
48		SH	685010	**Cynfal-fawr**	Nested once in last ten years.
49			685018	**Rhos-fach**	Alternate site to 48. Reported nesting early 1980's.
50	146	SN	675427	**Afon Cothi**	An old site, now used again.
51	160	SN	973397	**Cefn Merthyr Cynog**	An old site and reported as trying to nest in 1981.
52			984385	**Brestbailey**	Salter said Kite nested here in 1891. Another nest built since 1959 but it was left before eggs were laid.
53			950320	**Camnant**	Kite nested here until 1900. Two woods here, both having been used in recent years. Henbane says he took the young in 1801.
54			962282	**Cwm Camlais**	In 1906 Kites reared two young, but both were shot in the following December. In 1907 the eggs were taken on the well-known raid from London, when the party stormed the eyrie. Nested in various woods

Eyrie No:	Map		O.S.Ref	Name	Observations
					of the Mynydd Illtyd.
55			992328	**Mynydd Aberyscir**	An older site, not being used in recent years, but a possible future wood.
56			930380	**Blaendyryn**	Frequently used site. C/2 in 1979.
57	125	SH	898179	**Coed Cae**	A new site. C/3 in 1987. In Pine tree.
58			913188	**Cwm Pen-y-gelli**	Kites over this wood in 1987. A different pair from **57**
59	147	SN	906533	**Coed Pen-y-ceulan**	A long-established Kite site. Visited by Walpole-Bond in 1902. Nests here frequently. O.G. Pike photographed the nest on 16th March 1902 and the photo is in Walpole-Bond's book Bird Life in Wild Wales, page 264.
60	160	SO	015298	**Fenni Wood**	Scottish keeper at Frwdgrech shot five Kites in one winter around 1889-90. About 1883 a pair nested regularly in Fenni Wood. (J.H.Salter's research).
61		SN	915298	**Yr Allt**	Very old site. Nested regularly until about 1890. Vaughan Powell knew five pairs breeding within five miles of Sennybridge.
62		SO	046387	**Llaneglwys Wood**	An old Oak wood, but now surrounded by fir plantations. Kites seen here recently. Nested 1896 to c. 1899.
63			006395	**Pont-maen-Ddu**	Frequently nested here up to 1896. Has nested since at least once, but wood now much felled.
64		SN	998408	**Dan-yr-Allt**	Adjacent to site 35 and used for nesting in the late 1890's.
65	161	SO	113408	**Stockley Wood**	Nested here in the 1890's but the A 470(T) now precludes all nesting.

Eyrie No:	Map		O.S.Ref	Name	Observations
66	146	SN	634409	**Maes-troyddyn-fawr**	C/2 taken circa 1972. In frequented Kite area.
67	147	SO	035528	**Wellfield/ Cefndyrys**	Mentioned as an old site by Walpole-Bond and others. Nested until 1862.
68	135	SN	635797	**Rhiwarthen Isaf**	A new site. C/3 in 1985.
69	147	SN	843512	**Esgair Nant-y-Brain**	Site used in the come back of the Kite in the 1950's.
70		SN	897518	**Nant Annell**	Nested in 1985. Ornithologists chased off by man on horse before clutch size could be ascertained. Alternate site to Coed Caeper-corn.
71	147	SN	901540	**Llednant**	An extension of the woods of site 59. Used in many years lately.
72			957576	**Cwm Chwefru**	Bond mentions that this site was in a large Oak, opposite the farm. This was the old Cwm Chwefru Farm and not the new building at the end of the valley. Kite seen 1985 sitting in lower half of wood but workers felling trees precluded a proper examination. This felling must have put a stop to nesting that year. Salter says Kite bred for ten or twelve years from circa 1890. Only one young was allowed to leave the nest in all this time. Eggs taken after dark by a well-known egger in 1901. A bird seen by farmer on 24th March 1904.
73			988579	**Estyn Brook**	Two Kites prospecting in 1986.
74	124	SH	662197	**Uwch Mynydd**	Used within the last three years, with alternate site at 685197.
75	147	SN	716638	**Bronmwyn**	A modern site. Used most years. Man with shot-gun on guard.

Eyrie No:	Map		O.S.Ref	Name	Observations
76			940680	Rhydoldog	H.E. Forest informed J.H. Salter that one or more pairs nested here and at Treheslog up to 1868, when most of the trees were cut down. Reported to have tried to nest on the rocks, which would have been the only rock-nesting pair of Kites in Wales.
77	135	SN	640743	Llidiardau	Nested once in 1985. In range of trees behind the house.
78			660785	Bryngwyn	Another frequently used site along the Afon Rheidol.
79			725778	Afon Rheidol	Another nesting site in this much-used valley.
80	147	SN	780733	Cefn Coch/ Nant gau	In group of three Ash trees in the Cwmystwyth Kite area. 1987. C/3.
81	135	SN	893844	Glynhaffren	An old Oak wood. Reported nest in 1969 but Dawson could not find it.
82	146	SN	590413	Rhyd-y-mwyn	C/2 taken by Norman Gilroy on 17th April 1906. Few Oaks there now but suitable territory for Kites.
83	161	SO	123435	Craig Wood	A keeper yclept Sharp killed around thirty Kites during the years 1872 to 1902 in this area. The Craig Wood was the main area for a Kite's nest in those years.
84	160	SN	992284	Aberbran-fawr	A pair of Kites usually nested in this large wood, according to J.H. Salter, around the years 1890.
85	147	SN	856535	Ty-Mawr	Nested in round tall wood in 1988. Ravens also used this wood.
86		SO	062431	Fron Wood	Another old site in the Gwenddwr area.

Eyrie No:	Map	O.S.Ref		Name	Observations
87		**SN**	**709742**	**Nant Cwm-Newydion**	Nested 1986-87. The alternative site to Glandwgan No: 13.
88	**146**	**SN**	**585409**	**Wern**	Pair of Kites here in 1985. This is adjacent to Gilroy's site Rhyd-y-mwyn 82.
89			**630390**	**Pen-y-dolau**	One, possibly two Kites here in March 1983.
90			**666426**	**Cwm-Einion-fawr**	One Kite around here in Winter until March 18th 1985. Good territory. An alternate site to Nos 40 and 41. Nest usually in south part of wood towards Cefnceidryn.
92			**615307**	**Cwmcerrig**	In good Kite area, suitable for future use.
93			**445397**	**Llanfihangel-ar-Arth**	Another site for future use.
94	**135**	**SN**	**706960**	**Dynyn**	Birds around an empty nest in early April 1986. Probably eggs taken. May not have laid and moved elsewhere.
95		**SH**	**653043**	**Dolgach Falls**	Birds seen may have been the Pen Dinas pair.
96			**745029**	**Bron-y-aur**	In the new colonisation area along the Dovey valley. At least six pairs of Kites within five miles.
97			**792047**	**Coed-ddol**	C/1 taken in 1984 before full clutch laid. Kite usually continue laying if first egg taken, as stated by Walpole-Bond, but did not in this case.
98	**136**	**SN**	**997835**	**Moelfre**	Kites flying around here in 1981.
99	**160**	**SO**	**085238**	**Pant-llefrith**	A possible Kite site for future use.
100	**125**	**SH**	**833318**	**Coed Dolfudr**	
101		**SJ**	**012270**	**Blaen-y-cwm**	These three sites are in wild area.
102			**073293**	**Tan-y-pistyll**	A possible future use.
103	**147**	**SN**	**886526**	**Nant Bryn**	C/2 in 1978.
104			**887532**	**Bryn**	A possible alternative to 103.

Eyrie No:	Map		O.S.Ref	Name	Observations
105			**968718**	**Gilfach**	A suitable wood in Kite area, but not used in recent years.
106			**840455**	**Llethr-ddu**	Not inspected personally. Birds reported.
107			**893548**	**Lan**	Old site but few Oaks now left.
108		**SO**	**086608**	**Cwm-Brith-Bank**	Nested in the 1930's. Very steep. Now turned into a picnic area so unlikely to be used again.
109		**SN**	**895639**	**Rhosygelinen**	Old site but too many firs now. Birds moved to Glannau, No: 7.
110			**912626**	**Nant Gwyllt**	An old site now covered in conifers. Henbane took birds from here, but the actual site is beneath the waters of the Garreg Ddu Reservoir.
111	**146**	**SN**	**741413**	**Bedw Bach**	Walpole-Bond found a Kite's nest here and it was photographed by Oliver G. Pike on 19th May 1903 when it contained a c/2. Most Oaks now cut down and replaced by pines.
112	**147**	**SO**	**077428**	**Pant-y-celin**	One of the woods in the Erwood area used in the 1890's.
113			**088434**	**Erwood**	Another 1890 nesting site of the Kite.
114	**160**	**SN**	**785237**	**Blaensawdde**	A new Kite area, first nested 1978, although very probably used in the last century.
115			**763230**	**Panmaen**	Another altenative to **114**
116			**755220**	**Lan**	Yet another site. C/3's usually.
117			**750243**	**Twynllanan**	C/2 in 1979.
118			**783287**	**Sornddu**	An alternative site to Gwydre No. 40.
119	**147**	**SN**	**830581**	**Bryn Crwn**	C/3 taken by Norman Gilroy 1907. The original Oaks have now all gone and the area is a pine forest.
120	**124**	**SH**	**721235**	**Dolmelyllin**	Two young were hatched in 1980.

Eyrie No:	Map	O.S.Ref	Name	Observations
121		714245	**Gwyddyn**	Alternative site to No: 120.
122		757215	**Gelli**	Birds seen here. Probably from above two sites.
123		842233	**Foel Ddu**	One bird here in 1981. Possible future nesting site.
124		758104	**Craig Hen-gae**	Another suitable nesting wood.
125		676160	**Ty'n llidiart**	A possible site. Kites increasing in this area north of Machynlleth.
126	148	SO 106562	**Gwernfach**	Pair of birds diving into wood on 8th April 1976. Returned on 17th but some Rat had lifted the eggs.
127		107575	**Garnffawr**	Alternate site to 126.
128		121526	**Penarth Wood**	One Kite reported here in winter of 1979.
129		179512	**Upper Glasnant**	Bird flew towards this wood from the south-east. 1982. Perfect feeding areas on Glascwm Hill and Allt Dderw.
130		201675	**Neuadd**	Reported nesting in circa 1979. Woman in the Severn Arms, Penybont was asking persons if they had 'seen our nesting Kites?'.
131		126733	**Crosscynon**	A suitable wood near Llanbister. Bird here in 1979 could have been one from No: 130.
132	145	SN 325247	**Cil-hir-uchaf**	Out of the main area, but a bird seen here could mean future nesting.
133		323238	**Pen-Caerau-bach**	Ditto 132
134		355295	**Wenallt**	Ditto 132
135		362285	**Allt Dryscol-goch**	A very suitable wood Ditto 132
136	146	SN 485262	**Yspitty-Evan**	Small wood. Kite seen here 1987.
137	161	SO 114398	**Fforest**	Nested here in the 1890's, but road A 470(T) now precludes birds. Kites were seen in the Boughrood area and nested between here and Erwood in the 19th Century.

The following sites are on Map 160. They are all in the Brecknock District of the Mynydd Eppynt. All have had Kites nesting in them within the last one hundred years. About five pairs of Kites use this area for nesting each year. They are in addition to sites already listed above:

SN 902372 Aber-Cynog
906366 Tir-bach
906343 Garth-isaf
906327 Coed-Belli-du
906312 Tir Isaf
935350 Coed-y-carnau

SN 943337 Cwm-cynog
959385 Pant-Llwyn-on
969373 Yscir-fechan
982358 Fan-uchaf
988349 Graig
993363 Allt Ddu